The Tesla Files

*A Whistleblower, a leak, a fight for truth:
The inside story of Musk's empire*

SÖNKE IWERSEN AND MICHAEL VERFÜRDEN

MICHAEL JOSEPH

PENGUIN MICHAEL JOSEPH

UK | USA | Canada | Ireland | Australia
India | New Zealand | South Africa

Penguin Michael Joseph is part of the Penguin Random House group of companies
whose addresses can be found at global.penguinrandomhouse.com

Penguin Random House UK
One Embassy Gardens, 8 Viaduct Gardens, London SW11 7BW

penguin.co.uk

First published in Germany by C.H. Beck 2025
First published in Great Britain by Penguin Michael Joesph 2025
001

Copyright © Sönke Iwersen and Michael Verfürden, 2025

The moral right of the author has been asserted

Set in Garamond MT
Typeset by Couper Street Type Co.
Printed and bound in Great Britain by Clays Ltd, Elcograf S.p.A.

The authorized representative in the EEA is Penguin Random House Ireland,
Morrison Chambers, 32 Nassau Street, Dublin D02 YH68

A CIP catalogue record for this book is available from the British Library

HARDBACK ISBN: 978–0–241–79774–7
TRADE PAPERBACK ISBN: 978–1–405–98482–9

Penguin Random House is committed to a sustainable future
for our business, our readers and our planet. This book is made from
Forest Stewardship Council® certified paper

The Tesla Files

SÖNKE IWERSEN is an award-winning journalist who
has led the investigative team at Germany's top business
newspaper, *Handelsblatt*, since 2012. Known for uncovering
complex stories with precision, he has received numerous accolades, including four Watchdog Awards, the Kurt
Tucholsky Prize for Literary Journalism, and the German
Reporter Prize. Named Business Journalist of the Year,
Iwersen also teaches at the Georg von Holtzbrinck School
for Business Journalists and hosts the acclaimed podcast
Handelsblatt Crime, where he merges investigative rigor with
compelling storytelling.

MICHAEL VERFÜRDEN is a distinguished investigative journalist at *Handelsblatt*, renowned for his in-depth
reporting on topics such as abuse of power in the financial
sector, questionable real estate practices, and corporate-
targeted cyberattacks. His relentless investigation into the
Wirecard scandal unearthed secrets that even German
law enforcement agencies had missed. Verfürden teaches
research methods in cyberspace at the Georg von Holtzbrinck School for Business Journalists, blending traditional
investigative techniques with advanced digital tools.

CONTENTS

Trapped in Elon's Game

April 21, 2025

If Elon Musk had his way, this book wouldn't exist. When we sent him our questions about the Tesla Files in May 2023, the reply came straight from his legal department. We weren't allowed to have these files, a Tesla lawyer informed us, demanding their deletion. Tesla announced legal action, stating: 'As you know, use of illegally obtained data for media reporting is not allowed absent exceptional circumstances.'

Perfect, we thought – exceptional circumstances are a given. The world's richest man, the most valuable car manufacturer, the largest private space company – the story of Elon Musk is an endless chain of superlatives. It couldn't possibly be more exceptional.

For a long time, we never imagined belonging to a very small circle of journalists demystifying Tesla and, particularly, Elon Musk. Then, the whistleblower Lukasz Krupski contacted us.

His call turned Musk into the center of our lives, a constant presence. Since then, we have been exchanging countless messages about him, checking the latest news on Musk and Tesla early in the morning and late at night. When our alarms go off, our smartphone screens are already filled with his overnight posts on his platform, X. We sometimes asked ourselves how he still finds the time to run his companies between hours of video-gaming and his fight against what he calls the 'woke virus.'

On October 25, 2024, Musk casually uploads a video on X showing himself playing Diablo, an online role-playing game. In the background, a SpaceX team briefs him about issues with an orbital test flight. 'I want to be really up-front about scary shit that happened,' says an engineer. 'We had a misconfigured spin-gas abort that didn't have quite the right ramp-up time for bringing up spin pressure. And we were one second away from that tripping and telling the rocket to abort and try to crash into the ground next to the tower.'

Musk, who has security clearance due to his cooperation with the U.S. government, continues slashing skeletons with unnerving calm, replying without pausing his game: 'Wow, yikes!'

This near-catastrophe and Musk's relaxed approach to it barely gets any coverage. In the fall of 2024, Musk generates headlines at such a furious pace that journalists can hardly keep up. He produces news incessantly – news no other CEO could possibly survive.

All of a sudden, Musk has stepped into politics, not just as a donor, but as a contender. When we began our research for the Tesla Files, we thought the global public's interest in Musk had peaked. Now Musk – the protagonist of this book – has become involved in an unprecedented U.S. presidential campaign.

On the one hand, we welcomed this development – journalists would call it reporter's luck. On the other hand, we are not just reporters, but also citizens. Imagining Musk as a political decision-maker makes our skin crawl. The man is locked in an ongoing feud with regulatory agencies, but is now on track to lead the Department of Government Efficiency, an anti-bureaucratic bureaucracy. Donald Trump has suggested that this appointment is the modern-day equivalent of the Manhattan Project – the secret U.S. research program during World War II that led to the creation of the atomic bomb. It culminated on August 6, 1945, in Hiroshima, with a deadly demonstration of power that changed the world forever.

Today, the enemy isn't the alliance of Nazi Germany and Imperial Japan. Now, the enemy is anyone who thinks differently from Musk. Employees, business partners, and family members describe him as prone to outbursts of rage. His paranoid fears and conspiracy theories are blasted to millions via X, his own social media empire. The cost of his greed for success and recognition is paid, above all, by his employees. We've studied Musk's philosophy – his habit of minimizing present-day risks in favor of whatever the future might hold. In his world, the end justifies the means.

If the future looks revolutionary, today's damage is negotiable.

Even dropping an atomic bomb could be justified.

At times, while writing, we've felt that Musk approaches life as if it were a video game – with himself as the main character and everyone else merely extras. A game obsessed with becoming faster, better, stronger, where losses don't matter because progressing to the next level forgives every-thing – as long as you win. This book serves as a reminder that, in life, there are no automatic save-points. Levels can't be repeated indefinitely. And some mistakes can't be undone.

That we were able to write it, we owe to many people – first and foremost Lukasz Krupski. The trust this whistleblower placed in us was a rare gift. It took extra-ordinary courage to stand up to a hyper-rich, aggressive, and lawsuit-happy Elon Musk. Krupski has paid the price. His life has been stuck in a Tesla loop since 2019. And yet, he helped us tell this story.

Many others followed his lead. This book is built on the experiences, accounts, and information of numer-ous people who spoke to us directly and gave us access to internal matters. Every single one of them helped shed light on the Musk system. Many took on significant personal and legal risks. Some relived the worst moments of their lives for us.

This book has its origins in our first major exposé, in the German business newspaper *Handelsblatt* in May 2023, and the many articles that followed. But we wanted more for this project. Not only did we dig even deeper into the Tesla Files, uncover connections, build new relationships and bring to light secrets that had remained hidden, we also we wanted to take our readers behind the scenes of journalism itself. This book doesn't just show what's happening at Tesla – it also tells the story behind the story. It reveals the challenges of an investigation like this one: the confidential editorial meetings, the pressure, the push and pull. We want to show what it's like for reporters to swing between excitement and doubt – always aware that on the other side stands the richest man in the world.

For both of us, it's the first time we've written not just about our subjects and other people, but about ourselves, too. At *Handelsblatt*, journalists are taught from day one to take a step back from their own story. Shifting perspective – reflecting on our own work and our own role – was a challenge.

We regret to say we have no comment from Tesla or Musk to offer. Neither the company nor its major shareholder and CEO responded to our questions.

CHAPTER I

An Anonymous Caller

This particular Friday starts off gently enough. A conference has been canceled, and we recorded today's planned episode of the *Handelsblatt Crime* podcast yesterday. But the calm is shattered when a Microsoft Teams call pops up on my [Sönke Iwersen's] desktop.

Odd. The photo in the window shows Martin Kölling, our correspondent in Tokyo. I haven't spoken to him in months. What could he be calling about?

Martin gets straight to the point: A stranger called, claiming to work for Tesla in Norway – and warning of serious problems inside the company. Would I be willing to speak with the anonymous caller?

Of course I would. I'm not a car fanatic, but few companies in the world have the kind of magnetic appeal Tesla does. When Herbert Diess was still CEO of the

German car giant Volkswagen, he once invited Elon Musk to speak in front of two hundred top executives. I can't recall any comparable case in which a corporate leader so willingly threw himself at the feet of a competitor. Diess was practically enamored with Musk. Tesla, he said, worked faster, was more innovative, manufactured more cheaply – in short: it was simply better.

These days, Musk is practically the pop star of the global economy. A genius who founded his first company in his early twenties and was worth over $100 million ten years later.

What hasn't been written about Musk? How work-obsessed he is. How unpretentious. That he doesn't own a house and sleeps on friends' couches – or simply on the floor of one of his factories. That he plays by no rules. Since buying the short-messaging platform Twitter, Musk has been in the headlines constantly.

I would have loved to meet him at some point, but for German journalists that was virtually impossible at the time. We hadn't even managed to establish contact inside his German plant.

Today is November 4, 2022. For eight months now, Tesla has been building cars in Grünheide, near Berlin. It's Musk's only factory in Europe, and according to the company, the most advanced in the world. The building permit was rushed through in an expedited process. Prospective employees had to sign confidentiality agreements during their interviews. No one is talking to outsiders. The company is a black box.

And now, out of the blue, an insider comes forward. I'm skeptical. I've been leading the investigative team at *Handelsblatt* for ten years now. Again and again, people have approached me promising incredible scoops. On closer inspection, most of these supposed whistleblowers turned out to be pursuing their own agendas. Often, all sense of proportion goes out the window. A bank allegedly miscalculates a loan? A banking scandal. An insurance company refuses to reimburse a specific expense? An insurance scandal. Lose a court case? A justice scandal! There's no limit to how far personal grievances can be inflated.

And yet, I've learned this in my work: never turn anyone away. Always listen. In journalism, the most unlikely tips can turn into real stories.

In 2011, I heard a wild story involving Ergo Insurance, then widely known because of its massive marketing campaign. Independent agents from Hamburg-Mannheimer, a subsidiary of Ergo, came to tell me about a dispute with their employer. They were mainly fighting over severance packages for their many years of service. As I later learned, the men had already tried offering their story to other media outlets but without success. I wasn't all that enthusiastic either – until one of them mentioned, in passing, an event in Budapest 'that Ergo definitely wouldn't want made public.'

Suddenly, I was intrigued. The men exchanged odd glances – it seemed the Budapest topic hadn't been agreed on beforehand. I didn't let it go, and after a while I had pieced the story together. Hamburg-Mannheimer had announced a competition and invited its top agents on a pleasure trip to Budapest – a meticulously organized open-air orgy in the historic Gellért Baths.

The organizers had hired hostesses and prostitutes, marking them with color-coded wristbands, and stamped the women's forearms after each sexual service, for billing purposes. In the internal magazine for the sales team, the trip was praised as 'legendary' and 'a hell of a lot of fun.'

When *Handelsblatt* printed my article, it sent a shockwave through the entire industry. The trip caught the attention of police and prosecutors, was picked up by media outlets around the world, and in 2014 was even featured in the exhibition *Shameless? Changing Sexual Morality Through the Ages* at the House of History in Bonn, Germany's former capitol.

Since then, it has become customary at *Handelsblatt* for 'special' stories to land on my desk – or for special callers to be forwarded to my number.

As I write these lines, it's happening again. I've taken the day off to work on this book. The phone rings. It's a number from the United States.

At first, I think: waste of time. The caller is complaining about an article that just went live on the *Handelsblatt* website, about a Munich art gallery that has filed for insolvency.

The woman wanted to speak to the author. But he's a freelancer and doesn't have an office number here. So the receptionist has simply patched her through to me.

After a few minutes, I realize this might be worth it. The caller is an American art collector who's been in a months-long dispute with the gallery. She feels deceived – defrauded, even. The woman is elderly and tends to ramble. The conversation goes in circles until she suddenly mentions that she flat-out told the gallery owner she was a fraud. The owner buckled and supposedly admitted that her father had embezzled money from clients. Now, she doesn't know what to do.

I think it over. It's Friday afternoon. Maybe we could still reach a spokesperson at the public prosecutor's office today. On the other hand, it might be better to first find a few more customers who also had problems with the gallery. An investigation begins.

The Data Deluge

Back to Tesla. The anonymous caller who reaches out on November 4, 2022, is particularly cautious. He withholds not only his name but insists on communicating exclusively via the encrypted messaging app Threema. I know the Swiss service as something like the gold standard for privacy advocates.

Threema offers full end-to-end encryption for all messages, calls, and files. Users can use it anonymously, as no phone number or email address is required to register. Instead, they receive a randomly generated Threema ID. The company stores as little data as possible on its servers. Messages are deleted upon delivery, and no metadata – information on who communicated with whom and when – is logged. Threema has published its source code, so any expert can audit the software and verify that there are no backdoors or security flaws.

As cautious as the Tesla informant is about data security, he's equally eager to share information. Just 12 minutes after I send Martin my Threema ID, my phone vibrates. 'Hello Sönke, I have got your ID from Martin. Am I writing to the right person?'

I reply, 'Yes, that's me! I'm looking forward to your info!' Let's see what happens next. Only ten minutes pass. His next message reads: 'Okay, here are the termination spreadsheets.'

He's referring to the conversation he had with my colleague Martin. Allegedly, he was able to pull sensitive data about all employees worldwide from Tesla's internal IT system. He sends multiple files with his message. The first is an Excel sheet with more than 5,000 people Tesla has laid off. That's already a huge deal. Former employees can be excellent sources for journalists – especially if they were fired. But the termination table is only the beginning. It gets better. Much better.

Seconds later, another file arrives. And then another. The third one is the jackpot. What strikes me first is the size. This Excel file contains more than 30 megabytes. Spreadsheets typically don't take up that much space. I've never seen an XLSX file this large. How much data must be in here? I notice it's zipped – compressed. I have to unzip it to open it, and then it gets even bigger. In a follow-up message, the informant sends the password. I click on the file: 'StockAdmin Active Employee Headcount.' When it opens, I can hardly believe my eyes.

What I see appears to be HR data from Tesla. Column B lists first names, Column C last names, followed by hire dates, company email addresses, and personal addresses. Birthdates are in Column Z, right next to social security numbers. I click around, scroll here and there, then press and hold the Page Down key. Strangely, the scrollbar barely moves. I press it again. Ten seconds. Twenty seconds. Still no end in sight.

The amount of data is staggering. When I finally hit Ctrl+End to jump to the bottom of the document, I land on row 72,999. Just like that, I have access to email addresses, phone numbers, and more for over 70,000 Tesla employees. I open the next document. This one includes salary data. Another one lists reasons for dismissal. Some tables have more than 80,000 rows. Most files refer to current employees, some to former ones. In total, I'm looking at more than 100,000 individuals. In the U.S., in China, in Germany – everywhere.

I lean back and think. Can this be real? Data security has been a top priority in the corporate world for years. Companies run entire departments dedicated to managing and training people on how to handle sensitive data – whether it's about employees, customers, or business partners. The idea that someone at Tesla can simply pull an Excel file with millions of records from an internal system seems utterly absurd. The damage hackers could do with this kind of data is unimaginable.

In the U.S., if you know someone's name, address, and social security number, you have all you need for identity theft. With that information, criminals can apply for loans or credit cards in the victim's name and run up massive debts. They could also open bank accounts to launder money. Another common case is using stolen identity to access healthcare services – which leads to financial damages and mix-ups in the victim's medical records.

Preventing identity theft requires special care in handling personal data. That applies not only to individuals, but also and particularly to companies that collect and store such information about their employees.

All of this, I'm sure, must be known at Tesla. Its CEO Elon Musk sees his company as a tech firm, not a mere car

manufacturer. Processing and protecting data should therefore be central to everything the company does. It simply
isn't fathomable that an untold amount of sensitive information like this is floating around the internal network.

And yet here it is, staring at me from my screen. I ask the
informant: 'The documents were available to everyone?'

His reply arrives two minutes later: 'Yes, the documents
were attached to tickets that are not protected. Anyone can
search for them (ticket IDs) or for specific keywords like
"termination," "headcount master." I tested this with one
employee who I trusted and he also could access it.'

I show a bit of skepticism. 'That's highly unusual,' I write.

'It is. And very dangerous,' he replies. 'Although they
have monitoring apps on my laptop, no one reached out
to me from the IT department or Security. They (HR and
management) did not even ask for my laptop when I left
the company.'

It's just past 4:30 p.m. now. The unknown source sends
more files: details on the Fremont factory in the U.S.,
information on stock option plans for employees, even
files containing medical data. I could spend hours on this
– but in a few minutes, my daughter's daycare will close.
Thankfully, it's in the same building as my office. I can still
make it. I say goodbye to the informant and tell him I'll go
through everything carefully over the weekend.

Time to hurry. I shut down the computer. Melinda is
waiting downstairs. She's two years old. Shoes on, jacket

– she protests. It's unusually warm for November. We walk through the courtyard at *Handelsblatt*, then re-enter the building to get to the garage. Melinda stops when she sees Pepper, the robot that greets visitors at the *Handelsblatt* reception. She babbles in her baby talk to the white machine. Her favorite part is when I scratch Pepper behind his ear and he laughs.

Saturday is for soccer. My younger son has a match at noon, the older one three hours later. I sit behind the goal with my camera and take photos for the team – it's my job as a soccer dad. Today, Brandon's under-11 team wins 5:1. Dylan's under-15 team even wins 7:1. The boys are in great spirits.

That evening, the anonymous source checks in. 'Hi again, Sönke, I see that the last message did not go through. I have noticed that I did not send you one of the most important files in "Stock admin" zip archive. This one has probably the most personal information. Sending it now. Goodnight.'

Project T.

Back at the office, I begin to take a closer look at the files. The first problem crops up: some of the documents won't open. I call Michael Verfürden from the office next door and ask him to come over. He's been part of the investigative team for two years now and, alongside his journalistic instincts, brings a rare talent for data and technical detail.

For months, we've worked together on a series of articles about Wirecard. Once hailed as Germany's tech miracle, the company collapsed in one of Europe's biggest fraud scandals. The case shook the markets and exposed deep flaws in the country's financial oversight.

Michael is more than 20 years my junior, but he already shows all the right traits for this job. He's sharp, quick on his feet, and precise when it counts. He's drawn to big stories, but wise enough to question them. When I tell him about Tesla, he raises an eyebrow. I ask him to sit down at my desk and take a look. He scrolls. Then he stops.

Eyes fixed on the screen, he leans back and asks, 'Can this be real?'

One way to find out. We start working on the files but soon hit a wall when our Microsoft machines can't open all of them. The anonymous source uses a Mac, and some of the documents are buried so deep in subfolders that Windows loses track. Certain file types don't open at all. Eventually, we'll buy a Mac – just for this Tesla investigation.

Throughout the day, I continue exchanging messages with the whistleblower. He tells me he was once a victim of identity theft himself, years ago, and is deeply concerned about how carelessly Tesla handles the personal data of its employees. 'This info can be accessed by employees around the world. [. . .] Our adversaries, for example Russian and Chinese intelligence agencies, might be keen to have access to this kind of data and use it for their own purposes. We don't know what connections employees might have. Especially since 20% of the headcount is "recycled" every year. That means that access to this information has more and more people.'

I jot down a note. Can it really be that Tesla replaces 20 percent of its workforce every year? That alone would be worth an article in *Handelsblatt*. At Volkswagen, the annual turnover rate is just 1 percent. In companies where a high level of technical expertise is required, the goal is usually to keep employees for the long haul. Training takes time, and no one wants hard-earned know-how to walk out the door en masse. So why should it be different at Tesla?

I ask the anonymous source for his view – on Musk, on Tesla, and on the significance of what he's uncovered. It pours out of him. 'Musk has already harmed a lot of people, exploited them, deceived them. He puts dangerous software and hardware on the roads. I'd like to minimize any future harm to his customers and employees. [. . .] I saved his business by stopping the car fire in the delivery/ expo center in 2019, and he, in return, destroyed my life. They invented lies, discriminated against me, destroyed my career, finances, and health. But I loved my job and was committed to our company. People should know who he is and what kind of people he surrounds himself with. I hope that this sort of explains my point of view.'

Wow. I've never encountered a whistleblower like this. Much of what he writes, I don't yet understand. A fire? Saved the company? We'll have to verify all of it carefully. What a story this would be – if it's true.

It's nearly 10 p.m. when we exchange our final message of the day. Tomorrow, I need to talk to Martin Murphy, who has co-led the investigative team at *Handelsblatt* with me since 2021. And I should probably let Sebastian Matthes, our editor-in-chief, know as well.

Martin is thrilled. He's reported on the auto industry for many years and is, among other things, a true insider when it comes to Volkswagen. In 2019, we jointly received the Wächterpreis award for our reporting on the Diesel-gate affair. The watchdog award is given to journalists who uncover wrongdoing. Martin knows the awe with which German car executives revere Tesla. And he knows how closed off the U.S. company is. Martin has many contacts among German and international industry reporters – none of whom have ever managed to get Tesla insiders to talk, especially not to journalists. And now *Handelsblatt* might crack open this tightly sealed oyster? A sensational prospect.

We decide from the start to bring the entire investigative team into the Tesla project. Michael is already in the loop. A special role goes to Lars-Marten Nagel. Ten years ago, he worked as a data editor for the dpa news agency and has been part of our team since 2017. He teaches data journalism and research at journalism schools. Structuring and analyzing large datasets is his specialty.

As we present the Tesla topic in a Teams meeting, the whole department kicks into gear. René Bender and Volker Votsmeier – also award-winning reporters for their Dieselgate coverage – are on board from day one. Our plan: to discreetly test a few of the whistleblower's files. Let's see what happens when we start cross-checking the information from the Excel files with public profiles on LinkedIn, the German Xing, Facebook, and other social media platforms . . .

Next, I head to the sixth floor to see the editor-in-chief. Sebastian Matthes has led *Handelsblatt* since 2020. He has a strong interest in new technologies and reported early in his career on the green transformation of the economy. Tesla is one of the brightest stars on his radar.

Sebastian listens intently as I tell him about the unexpected insider who contacted us – about the data of 100,000 employees now visible on my screen. If the information is accurate, we could immediately report on the executives closest to Elon Musk. On Tesla's pay structure. We could find the right contact for any topic worldwide – all thanks to a single file.

Sebastian's first response: 'I don't think the data is real.' He reacts just as Michael and I had. Of course every journalist craves a world-class scoop. But for one to simply fall into your lap? Unlikely. Tesla is a high-tech company, and Musk is a skilled programmer. None of us at *Handelsblatt* can quite imagine a global corporation, with a CEO like that, being so careless with data protection. Then again, the story sounds too good to ignore.

Peter Koppe shares that view. The head of our legal department is the go-to for the investigative team. Not a single one of our stories gets published without first being reviewed by our legal experts. We publish stories that ruffle feathers. If they were legally vulnerable, we'd have a problem. That's why we've long made it standard practice to involve Peter's expertise early on.

When I first tell Peter about the files, his excitement is short-lived. 'Assuming they're real,' he says, 'how can we be sure these files were actually accessible inside the company? Couldn't they have been stolen by a hacker?' His suggestion: Watch the whistleblower retrieve the data.

If only. The whistleblower insists on absolute anonymity. According to him, he raised concerns about misconduct at Tesla early on – and was bullied, demoted, and eventually fired for it. He doesn't want journalists to know his identity. He points to Musk's public vendettas against critics. He firmly declines any meeting.

I keep pushing. 'The files you've sent are very interesting,' I write. 'Are they the only ones that were this easily accessible? Or is there more? And could you have copied additional ones if you'd wanted to?' His response sounds almost incredible. In his estimation, ALL data at Tesla is unprotected.

'There are hundreds of thousands of tickets,' the whistleblower writes. 'A lot of them have sensitive, confidential and privileged information. [. . .] It concerns nearly all departments. When it comes to GDPR (employees' personal data), as soon as I discovered the terminations spreadsheet, I searched for more tickets like that to see if this is a pattern. And it was. I bet it's still happening. So, to answer your question, I think there is more.'

He gives examples. 'I had access to global payables. I could see all the bills filed to Tesla from various suppliers. Sometimes even there were PDFs from big legal companies.'

My God, I think. Invoices? Could we actually see the terms under which major German suppliers like Bosch or ZF are working with Tesla? That would be explosive material. And PDFs from big law firms – that sounds like contracts. The kind of information journalists rarely ever get their hands on.

A Reporter's Paradox

As our appetite for the data keeps growing, our fundamental problems remain unresolved. How can we know whether the information was freely accessible, rather than being hacked? I try a different approach: Could the whistle-blower share his communication with the U.S. authorities he first informed about the situation? I'm hoping to find clues there about the origin of the data.

He refuses. 'The thing is that you would have my name on file,' he writes. 'If someone hacks you I could risk retaliation. Especially if company goes down. Musk is evil.'

I slowly begin to realize the bind Anonymous is in. 'I don't have a lawyer now. I am unemployed,' he writes. 'I would have no means to defend myself. And we barely know each other.'

He's right. It's the paradox journalists live with. We're constantly trying to draw out secrets from people we hardly know. On the other hand, anyone who approaches a journalist with a sensitive issue has already made the decision to open up to a stranger. I just have to be patient. Make it clear to him that, as a reporter, I can't just write things down on request. If someone wants a grievance brought to light, I need information I can verify.

Apparently, he's come to the same conclusion. Less than two hours after laying out his reasons for caution, I have his submission to the SEC – America's Securities and Exchange Commission – on my screen. Two documents, more than 30 pages. He outlines security deficiencies, issues in accounting, production problems, and a range of other concerns. Some of it seems a little too detailed. But we're still at the very beginning of the investigation – and in my mind, no journalist should ever complain about having too much information.

It gets better. My source suddenly has the same idea as our in-house counsel, Peter Koppe. He can no longer access the Tesla system himself, but he has a friend still working at the company – someone he trusts. What if the two of them recorded themselves searching for and retrieving data? Would that help *Handelsblatt* confirm the authenticity of the files and the lack of protection?

I accept immediately. I can work out the legal fine – points with Koppe later. There's no doubt that a demonstration like this would be a huge step forward. Anonymous says he's just sent a quick message to his friend. As soon as he hears back, we'll move forward.

Now might be a good time to bring up another delicate issue: what's the competitive landscape? Journalists are always in competition with each other – for news, for stories. The race is fierce enough when it's about domestic issues. Tesla is in the global spotlight. The obvious question: has Anonymous spoken with other journalists? Are we the only ones he's shared documents with?

The answer feels like a cold shower. No – he's already shared them with a reporter at a major U.S. news outlet. In January. I have to take a deep breath. In January?! What's that supposed to mean? Does this guy think *Handelsblatt* publishes large stories on topics that have been covered elsewhere ten months ago? Or did an American reporter review the data and find out something was wrong with it?

A few minutes go by, then he writes again. He says he was disappointed by the journalist. She had promised to write many articles on various Tesla-related topics. That didn't happen. And when he suggested disclosing some things in Europe, she refused. It 'prolonged everything,' the Tesla man writes. 'I could've spoken to someone trusted a long time ago. [. . .] What she wanted was to convince me to share publicly my personal story and exploit me as a source. Which is not OK.'

Now uncertainty is gnawing at him. 'I regret handing her this and the worst part is that I have no control over what she does with this sensitive information.'

I try to assess the situation. Tesla is a hard nut to crack, that much is clear. Can we really deploy significant investigative resources if we have to fear that the competition might beat us to it? Ten months is a lead that's nearly impossible to make up. Then again, ten months is awfully long for a head start. My hunch? The American isn't working on the story anymore.

Every now and then, journalists do miss out on major scoops. Before the *Boston Globe's* Spotlight team exposed the sexual abuse of children by Catholic priests, victims had spent years trying to get reporters' attention. NSA whistleblower Edward Snowden recounts in his memoir how he repeatedly failed to persuade journalist Glenn Greenwald to communicate over a secure channel. It was only through documentary filmmaker Laura Poitras that the two eventually connected. Greenwald very nearly passed up the biggest story of his life. I've had similar experiences – though never on this scale.

The anonymous source and I continue exchanging messages, dozens of them. He sends me links to articles published by our U.S. competitors. They're reassuring. The stories are interesting, but there's no blockbuster. No mention of the massive data leak, nothing about security vulnerabilities. One curious detail: the journalist had apparently passed along a select document from the Tesla insider to a German news magazine. An article appeared there as well, but nothing that would limit us at *Handelsblatt*.

Then the source makes a proposal. That morning, he has spoken with his former colleague about making a video. The colleague agreed to log into Tesla's IT system with him and record how they enter search terms, retrieve data, and download files.

The whistleblower's trust in me and my team seems to grow. When I propose having him speak with two colleagues from my investigative team, he agrees shortly afterward. The next day, he sends me the contact details of the two attorneys who assisted him in his labor dispute with Tesla. From a journalistic perspective, that's another step forward. Of course, lawyers speak on behalf of their clients. But being able to consult a professional legal opinion is a clear plus. I tell Michael and the rest of the crew the good news. Things are definitely moving in the right direction.

Later that same afternoon, my source and his friend sit down in a public library. They film themselves entering various search terms into Tesla's internal project management system, retrieving data with ease, and downloading files. I access the video via Dropbox; it's another milestone in our investigation. In just two and a half weeks, we've gone from an almost unbelievable tip from a stranger to concrete video evidence. If we keep the pace up . . .

I double-check: the source doesn't want us to tell his story, does he? He wants to appear only as a source, not a protagonist?

That's correct. 'I'm not ready for that kind of confrontation,' he explains. According to him, Tesla is not a normal employer, not a normal company. The leadership does not care about the truth, about doing the right thing. If he were to come forward, he says, he would become a target. 'They could just present me as a disgruntled employee, which could hurt the article,' he writes. 'I loved my work and the company until they attacked me and broke my life. It is probably rather more essential to focus on the issue than on a person.'

We agree that in any article he won't be identified as a former Tesla employee, but simply as a whistleblower. He also agrees to let us quote a few lines from his submissions to various authorities. That way, we can show readers the information is authentic.

The Whistleblower Turns Up the Heat

The entire *Handelsblatt* investigative team is buzzing with Tesla energy. Discoveries are constantly flying back and forth, usually followed by questions. I collect them and pass them on to our anonymous source. The steady stream of messages creates a strange sense of closeness. In these weeks, I exchange more thoughts with this man – whose name I still don't know – than with anyone else. We write at all hours of the day, sometimes even past midnight, and weekends go without saying. On Monday, November 28, there is something about Musk on his mind.

'I assume you did see the tweet about guns,' he writes in the evening. Elon Musk has just tweeted a photo of a nightstand: two guns, four cans of Diet Coke, and a portrait of George Washington. The caption: 'My bedside table.'

We talk about what it's supposed to mean. Neither of us understands what Musk is doing – or what's driving him. His behavior seems more irrational by the day. Can you imagine any other CEO claiming to sleep next to pistols?

Then I reach out to the whistleblower's lawyer. On the

phone, I only get the receptionist, so I follow up with an email. The next day, something strange happens.

'Hi Sönke, thanks for reaching my lawyer. I received an email from her today telling me that I will be invoiced for her time spent responding to you.'

I blink. 'Uh oh. How much money are we talking, then?' I ask. The answer: 2,468.75 Norwegian kroner. 210 Euro. Per hour.

'I have never met a lawyer who wanted to be paid for talking with me,' I write back. 'Does she realize how much publicity this could mean for her?' He replies: 'I am not sure if she is obliged to do that (request payment). To be honest, I am not exactly happy with them. When my contract with Tesla terminated, and there was time to act, they sort of abandoned me. They requested 500,000 NOK in advance for a lawsuit [about 45,000 Euro]. They knew that I could not afford that. I am still paying them back for the costs incurred during my employment.'

Good grief. I tell him we'll check at our end whether we can cover the cost. Then I bring up the timeline. The Tesla data contains multiple stories. There's no way to fit everything into a single article. So, we are looking at multiple stories, and each one of them has to be researched thoroughly, structured, and written. Then comes fact-checking and legal review. Even in the best-case scenario, we wouldn't be able to print the first article before mid-December. That means all subsequent stories would fall into the holiday season, when most people have only Christmas on their minds. From his point of view, is there any reason not to wait until early 2023 to begin publishing?

He's skeptical. 'There is a risk that once authorities move on Musk and this won't be news anymore as more dirt will come out. That's my feeling,' he writes. 'From what I see, Musk is damaging the brand and his name more nearly every day.'

We go over it again and again. In his view, the authorities could act very soon. At *Handelsblatt*, we come to a shared conclusion: we have no choice. We have to work as fast as possible – but without compromising quality or accuracy. What we don't know yet is that it will be many more months before even a single line goes to print.

I've spoken with Peter Koppe several times by now – our head of legal. He loves a story with bite. But when it comes to legal footing, he won't give an inch.

He seems cautiously pleased – the lawyer, the video. 'Good,' he tells me. 'But don't fool yourself. It's not enough.'

What if we imagine the worst-case scenario? That everything this unknown source is presenting to us has been planned and orchestrated from another side entirely? That the data and the video are a deepfake – created solely to drive Tesla's stock down? Every single percentage point at the stock exchange means billions here. Malicious actors could deliberately try to deceive journalists to plant damaging articles in the media. We need to rule that out. A video won't do it. And the lawyer can't prove the data is genuine either.

I make another attempt. 'I don't know how you feel, but it seems to me over the past few weeks we have come a long way in understanding each other,' I write to Norway. 'I wonder what you would think about me flying up there and spending half a day talking about all this. I think it could really help – though I do understand your reluctance.'

Then something happens that I've encountered a few times in this line of work. After weeks of firm refusals – no way he could agree to a meeting, not even a Teams call with a visible email address – suddenly a visit to Norway is no longer an issue.

'I don't mind, to be honest,' the anonymous source writes me, just two minutes after I suggest a visit. 'I could be easily identified by the documents I provided to you anyway.'

It's just what I needed to hear. It's 9:39 a.m. on December 14, 2022. We make plans to meet the following Tuesday – in six days' time. My destination will be the Norwegian port city of Drammen, 45 kilometers southeast of Oslo. The whistleblower who just days ago was shy and elusive suddenly turns into a tour guide, sending me train schedules and hotel tips.

The Christmas Scoop

On December 20, 2022, at 11:25 a.m., I send him what will be my last message from Germany for the time being: 'On the plane now. We are a bit delayed.' Two and a half hours later, I'm standing at the airport train station in Oslo. The ride to Drammen takes me through a picture-perfect pre-Christmas scene – snow everywhere, trees blanketed white. In Norway, you can still count on winter. I let the Tesla man know my arrival time.

At 3:30 p.m., I'm in my hotel room. Twenty minutes later, my phone buzzes: 'OK, I am downstairs. Blue coat, light-brown trousers and glasses.' A few minutes later, I take the elevator down to the lobby. And there he is. As I see him for the first time, I realize I hadn't once asked myself what he might look like – the man with the Tesla secrets. About 1.8 meters tall, mid-thirties, I'd estimate. He smiles politely, maybe a little unsure. We shake hands – and that's when I first learn his name: Lukasz.

We step out into the Norwegian cold. His car is just a few meters away – a massive Land Rover, perfect for the snow drifts ahead. From our hundreds of chat messages, I know Lukasz is into football. The World Cup in Qatar has just ended – Lionel Messi has finally secured the long-coveted world title. A perfect topic to warm up over.

Lukasz drives his Land Rover into a parking garage in downtown Drammen, and we settle on a restaurant with international cuisine.

Lukasz slips easily into storytelling. I encourage him to just start from the beginning. Where is he from? Why is he here now? He's happy to share. Grew up in Poland, traveled the world, became fascinated with technology early on – and eventually a devoted admirer of Elon Musk. In 2018, Lukasz moved to Norway for a job at his dream company. There, he experienced the greatest disappointment of his life.

He gave everything to Tesla, Lukasz says – even risked his health. He tells me about the incident at a 2019 auto show – one he had already mentioned during our early chats. Lukasz describes how he literally reached into flames to prevent a disaster. How, afterward, he received an email from Elon Musk himself.

For a brief moment, he thought the richest man in the world might actually be interested in his ideas – in his warnings that something was going wrong at Tesla.

'But I was wrong,' he tells me. 'The company spat me out like a contaminant.'

Time flies. We mostly talk about Tesla, but also about other things. I notice that Lukasz is a curious person. He asks a lot about *Handelsblatt*, wants to know what kinds of stories our investigative team has worked on. He listens intently as I tell him about my trip to Hong Kong, where I tracked down the people who hid Edward Snowden after his first interview.

Again and again, our conversation circles back to Tesla – and especially to Elon Musk. It quickly becomes clear that Lukasz knows far more about the Tesla CEO than I do. He's read the books, watched the documentaries, follows every post Musk makes on Twitter. Apparently, he's been doing this for years.

Lukasz seems to know every twist in the Twitter acquisition. He tells me who Musk has targeted, how he's gone after each critic, what tactics he used. How relentlessly he pursues anyone he labels an enemy.

It feels slightly surreal to hear Lukasz talk like this – especially knowing what we're here to do tonight.

Soon, it's time to go. Time for the acid test. We leave the restaurant and head back to the car. My tension rises. Will this mission succeed? Lukasz no longer has access to Tesla's IT system. The friend who filmed the library video with him has been hard to reach in recent days – he's sick. There's no guarantee that he'll be willing to open his laptop tonight for a German journalist. Still, I came here because I couldn't pass up the chance to meet Lukasz in person. Now I'm standing beside him in the parking garage, watching as he dials his friend.

Lukasz is Polish, his friend is Greek. The two of them speak English with each other. From what I can make out, the friend seems to have little interest in doing anything at all tonight – let alone showing a journalist Tesla's internal secrets. But Lukasz is persistent. After ten minutes, he ends the call and says: 'We can go over. He'll give us 20 minutes.'

It's really happening. The very task I came to Norway for is about to be fulfilled. We get back into the Land Rover and drive to Lukasz's friend's place. We park outside the building where he lives, and Lukasz sends him a quick message. A few minutes later, a man steps out onto the street. He's wearing a backpack.

We greet each other briefly. The man looks pale and feverish. He's coughing. Best to move fast. Back to the hotel. On the way to the meeting, we had discussed whether there might be a more secure place for what we're about to do. On the other hand, Lukasz has already come to terms with the fact that once the reporting begins, his identity will come to light sooner or later. The same goes for his friend. Over the past few months, the two of them have opened hundreds of so-called tickets in Tesla's IT system – without having any professional reason to do so. It's a wonder Tesla hasn't pressed charges against them already – or at the very least revoked their access.

The lobby at the Comfort Hotel Union Brygge is deserted this evening. As the three of us settle into a corner, the man at the front desk glances up briefly, seems to recognize me from check-in, and takes no further notice of our meeting. There's barely any guest traffic – we have privacy. The Greek pulls a laptop out of his backpack. He wakes the Tesla-issued device, fingers flying over the keyboard as if it's just another day at work – only it's not.

Lukasz sets up a second laptop on the table and powers it up. That's when I notice his last name: Krupski.

The two of them give me a brief tour of Tesla's digital landscape. The most important program is Jira, a project management system designed to improve team efficiency. I watch as the Greek opens the application. There's a search bar where employees can find the files relevant to them. But as already shown in the video, you can also find things that have nothing to do with your own field of work.

Lukasz's friend is, like him, a maintenance technician – right at the bottom of Tesla's corporate hierarchy. He opens a few files at random. Then I suggest we proceed methodically. First, I want to see if he can retrieve the files I already know – like the personal data of 100,000 Tesla employees.

No problem at all. I see the familiar Excel file 'Headcount Master'; it opens and can be copied. The same goes for other documents Lukasz had already sent me via Threema. That point seems settled. So what's next?

'Musk,' I say. And Tesla's IT system delivers. Hundreds of files show up under that search term. We narrow it down. We find contracts with his name, extensive correspondence

– even an invoice from the security firm that protects Musk on his travels.

I can hardly believe what I'm seeing. No matter what keyword we enter, the system spits out a flood of documents. Again and again, the first pages are stamped with words like 'confidential' and 'secret.'

There's no question: a technician at the lowest rung should never have access to any of this. And yet, no warnings appear. No access is denied. Nothing.

It's insane. What if I were an employee at a rival firm? Or a corporate spy? This data could change hands in seconds. In the wrong hands, it could cause real damage.

I marvel at the recklessness of it all. How can this be allowed to happen?

But of course, as a reporter, I'm more than thankful that it is. I'm a happy witness.

Jan Marsalek appears

It's been a little over an hour since we sat down in the lobby. By now, Lukasz's friend has already tripled the time he initially agreed to give us. But since he doesn't object, I simply pretend I don't see the clock in the bottom-right corner of the screen. How could I stop now, with the whole Tesla world open before me?

Lukasz asks what we should search for next. I hesitate. My main objective has already been fulfilled. So why not get creative? I spell out the German word '*Staatsanwaltschaft*' – public prosecutor's office.

What happens next is one of the strangest moments I've ever experienced in my career. I didn't give the word much thought when I suggested it. But in my experience, documents containing '*Staatsanwaltschaft*' tend to be explosive. I had no idea what a search for that term on the Norwegian site of an American company might turn up. But if I'd had any expectations, they would have been blown out of the water.

I skim through the results on the screen. The third one catches my eye. '20201223_703_St_2of_AnO_TKÜ_Tesladaten_081727' is the file name. I figure the first part indicates a date – December 23, 2020. *Tesladaten* is self-explanatory. The abbreviation *TKÜ* I know from previous investigations: it stands for telecommunications surveillance.

'This one – open it,' I say to Lukasz. Good thing I'm sitting down, or I would have fallen off my chair. 'Republic of Austria, Department of Justice, Public Prosecutor's Office Vienna,' reads the top of the first page. 'Order to provide information on communication data.' And then: 'Criminal case against Jan Marsalek.'

Am I dreaming? I read as fast as I can. The letter from the Vienna prosecutor is addressed to Tesla's European headquarters in Amsterdam. A prosecutor is requesting 'all stored geolocation data' for the Tesla Model S used by a former member of Austria's National Council in 2020. Authorities suspect the politician acted as a service provider for the disgraced German company Wirecard – and as an accomplice in Jan Marsalek's escape. The former Wirecard board member vanished following a multi-billion-euro fraud and has been wanted internationally ever since.

Lukasz and his friend stare blankly at the screen. I keep reading, transfixed. 'According to current reports by the Federal Criminal Police Office [. . .] there is a strong suspicion that the currently seconded BVT officer, Mag. Martin W., and other yet-to-be-identified employees of the BVT who worked freelance for Wirecard AG to verify the solvency of providers of pornographic websites,' reads the justification for the request to Tesla. Moreover, former parliamentarian Thomas S. is suspected of having helped Marsalek flee using his Tesla.

The demand to Tesla is sweeping. 'All stored GPS location data (historic vehicle events), especially "tracklog" data (lat, ion, speed, time, engine start/stop, parking positions and durations, door openings) is required.' Tesla is prohibited from informing any third party about the prosecutor's request due to confidentiality requirements.

I turn to Lukasz. 'Do you know what this is about?' I ask. He shrugs. He recognizes the name Wirecard, and Marsalek – his 'Wanted' posters are plastered across airports and train stations throughout Europe. Then he asks the obvious question: 'What does this have to do with Tesla?'

In this instant, it's clear to me: this can't be a fake. I chose the search term myself. No one planted the document as bait. That the word '*Staatsanwaltschaft*' came to mind was pure coincidence.

The Tesla data leak is real – and particularly serious. I've just read how the public prosecutor's office in Vienna instructed Tesla to keep the order concerning Marsalek's getaway vehicle confidential. And now, an employee from Tesla's lowest rank has simply pulled it out of the IT system – accessible to 100,000 people.

That should never have happened.

Mission accomplished. We try a few more search terms, but I can tell I'm testing the patience of Lukasz's friend. He had offered 20 minutes of help – nearly two hours have passed. I thank him warmly. Lukasz and his friend shut down their computers. We say our goodbyes. I return to my room. Time to make some notes.

Six weeks into this story, and tonight we crossed a threshold. We're not chasing a lead anymore. We're sitting on something explosive.

The Children's Café Handover

The next morning, I brief my colleagues at our head office in Düsseldorf. It's three days before Christmas – but our gift has already arrived.

We have the full name of our whistleblower, and that of his accomplice. Proof that ordinary technicians can access highly confidential files in Tesla's IT system. A remarkable discovery pulled straight from that very system.

And the day is just getting started.

'Good morning,' I text Lukasz just before 9 a.m. 'What are your plans today?' He doesn't have any. I seize the moment. 'Great. Would you like to come over at 10 and we can hash out more ideas?' He would.

At exactly ten, Lukasz arrives. We drive around Drammen for a while, then settle down at a café. I ask him how much data he actually has, and what state it's in. Several gigabytes, he says – mostly unsorted. He began digging for information when his dispute with Tesla started. In the beginning, it was just screenshots, fearing someone might notice if he accessed files that weren't his to see. But over time, he grew bolder.

We talk through the different angles I could pursue as a journalist. Data security is the obvious first step. But over dinner yesterday, Lukasz already mentioned several other topics. Issues with Tesla's Autopilot. Technical flaws in the vehicles.

Now he tells me about a file containing hundreds of customer complaints — and the absurd replies from Tesla employees.

This just keeps getting better, I think. There are still a few hours left before I need to head to the airport. Everything is going smoothly.

Then Lukasz suddenly says: 'Would you mind if we switched cafés?'

Huh. What's going on? I ask if everything's OK. He nods. 'It's just . . . people can see us from every angle here,' he says. 'I'd feel better somewhere else.'

Golden rule of journalism: if a source wants to go somewhere they can't be seen giving you something – go.

No sooner said than done. We get up, pay, and a few minutes later we're in a different café — though it looks more like a hall. Very spacious, with plenty of room to run around. Most of the guests have kids with them. While the parents sip their lattes, the children play with building blocks on the floor. Lukasz and I sit at a table along the edge.

No one pays any attention as he opens his laptop.

Again, he gives me a glimpse behind Tesla's facade. This time, it's excerpts he's already copied. He shows me Excel files listing various issues with Tesla vehicles. Photos from the auto show where he prevented a fire. The email he received from Elon Musk afterward.

Then Lukasz says, 'Did you bring a hard drive?'

I brought two. It's the very question I'd hoped for while packing for Norway. I open my suitcase and hand him one. Lukasz plugs it into his laptop.

'Which data do you want?' he asks.

What should I say? There's really only one answer.

'Well, since you're asking – everything you've got,' I reply.

He smiles. 'That's what I thought.'

And starts copying.

I watch the files transfer from one device to the other. One hundred megabytes. Five hundred. A few minutes in, it's several gigabytes – and according to the progress bar, we're not even halfway through.

What a scene. To my left: Norwegian children playing; some parents join in, others tap away on phones or laptops. To my right: Tesla files pouring onto my hard drive.

Meanwhile, Lukasz pulls more devices from his bag – USB sticks, SD cards, flash drives. He flips through them.

'Oh, here's an older folder,' he says. 'Mostly screenshots though.'

'No problem. I'll take it,' I say. A few clicks later, those files are flowing too.

When I stow the hard drive back in my suitcase, I'm carrying over 19 gigabytes of Tesla data.

Time is running out. We pack up in silence. Outside, the streets are slippery with snow and ice. I walk slowly, every step deliberate. There's too much at stake.

The train station isn't far. I buy my ticket, shake Lukasz's hand.

'Thanks for everything,' I say. 'See you soon.'

He nods. He looks relieved – and hopeful. For months, he'd been waiting for someone to take his data seriously. An authority. A journalist. The big bang never came. Now he's placing his trust in *Handelsblatt*. And in me.

The return trip is smooth. The train is on time. So is the plane. At security, no one asks about the devices in my suitcase.

At 4:15 p.m., I text Lukasz: 'Boarding is about to start. Thank you. And thanks again for all your time and guidance!'

He replies without hesitation: 'You're welcome. You know, anything to help. And thank you for taking this seriously.'

CHAPTER 2

Elon Musk, Superstar

When Vladimir Putin's troops invade Ukraine on February 24, 2022, the defenders are virtually blind. With a cyber-attack, the Russians have knocked out the satellites that provide the country with internet and telecommunications. Ukraine sends out a call for help. But it doesn't turn to the White House or to NATO. The Ukrainians appeal to Elon Musk.

Deputy Prime Minister Mykhailo Fedorov posts on Twitter, addressing the billionaire directly: 'While you try to colonize Mars – Russia try to occupy Ukraine! While your rockets successfully land from space – Russian rockets attack Ukrainian civil people! We ask you to provide Ukraine with Starlink stations and to address sane Russians to stand.'

Musk agrees – and just like that, he becomes a player in the war. His satellite internet service, Starlink, run by his space company SpaceX, turns into an existential factor in world politics. Musk now plays a role in determining the course of a war. He keeps repeating that, without his technology, Ukraine would have long since been overrun. But . . . he doesn't want to help for free.

In October 2022, Musk submits the bill for Starlink to the U.S. Department of Defense – even though no contract exists at the time. (An agreement has since been reached.) But Musk's unpredictability, and his repeated calls for a negotiated end to the war, are causing deep concern among politicians and military leaders. They know that a mere nod from Musk could once again plunge Ukraine into digital darkness.

This episode reveals the sheer scale of his power. By the end of 2022, Musk owns a fortune of more than $200 billion, making him the richest person on the planet. Next to Tesla, even Germany's largest carmaker, Volkswagen, is a dwarf in terms of market value. Since taking over the short-messaging platform Twitter, Musk can broadcast his views to over 100 million screens at once – or silence opinions he doesn't like. And yet, the multibillionaire is more than the sum of his companies' metrics.

To understand Musk's influence, one has to ask a simple question: when has a private citizen ever had as much impact on the outcome of a war, as Musk does in Ukraine?

When we begin our investigation, no one comes to mind – aside from Tony Stark, the comic-book billionaire in *Iron Man*. Both are visionary entrepreneurs. Both want to save the world. Both ignore the rules. And indeed, in 2010, Musk appeared in a cameo in *Iron Man 2*, playing himself. He was the inspiration behind the modern incarnation of the comic-book character.

Musk becomes the center of our reporting. Who exactly are we dealing with here? For years, we've followed his rise to the top ranks of global power only loosely. Now, he dominates our attention. We order every book about Musk we can find. We trade the latest articles about Tesla and his other ventures back and forth. Over lunch in the *Handelsblatt* cafeteria, we compete to one-up each other with absurd moments from interviews of his we've found on YouTube.

From the scale of his personal and entrepreneurial escapades, Elon Musk is one of the most entertaining figures we've ever researched. And from the scale of his power and influence, one of the most unnerving. Sometimes, the two sides blur.

Take the photo that shows Musk standing in front of an American flag, wielding a samurai sword. His eyes are focused, his arms tensed. When a user on his platform X shares the image, Musk replies: 'There is a large graveyard filled with my enemies. I do not wish to add to it, but will if given no choice. Those who pick fights with me do so at their own peril.'

Is Musk a madman, an eccentric, or the greatest genius of our time? And what should we prepare for, now that we're investigating the darker corners of his corporate empire? To understand what we're up against, it's worth looking back: how did Musk become the man he is today?

A Violent Beginning

Elon Musk was born on June 28, 1971, in Pretoria, South Africa. His father, Errol, a fast-moving engineer and businessman, was always chasing the next big deal. His mother, Maye, a dietitian and model, became his counter-weight. Their marriage was marked by conflict and violence. A prelude to Elon's own life.

In Walter Isaacson's biography of Musk, people who knew the family describe Errol as a man full of contra-dictions. Errol's father – Elon's paternal grandfather – was a cryptanalyst and intelligence officer who spent his days solving crossword puzzles and drinking. His mother, Cora, had left school at 14 to work in a factory.

Their son was hardworking. Errol studied mechanical engineering and later built hotels, shopping centers, and factories. On the side, he restored vintage cars and airplanes, involved himself in politics, and won a seat on Pretoria's city council. He trained as a pilot and teamed up with a businessman who owned an emerald mine in Zambia. Errol imported raw emeralds and had them cut in Johannesburg. Many people also offered him stolen stones to sell abroad. It was an illegal but profitable business.

Errol was mercurial. At times generous, then suddenly dark and cruel. In Isaacson's biography, Elon recalled evenings when his father forced him to sit for hours in a Pretoria casino, scribbling down numbers while Musk Sr. worked on a gambling system he believed to be unbeatable. His son was merely a tool.

Maye and Errol's marriage was fraught from the beginning. He proposed several times, but she didn't trust him. After graduating from college, Maye moved to Cape Town. Errol visited her there and proposed again. This time, she said yes.

Their relationship was steeped in cruelty. 'He would hit me when the kids were around,' Maye wrote in her 2019 memoir. She recalled Elon's sister Tosca and brother Kimbal – two and four years old at the time – crying in the corner, and 'Elon, who was five, would hit him on the backs of his knees to try to stop him.'

Elon was eight when his parents divorced. Maye moved with the children to the coastal city of Durban. Money was tight, and she had to work while the children looked after themselves – sometimes well, sometimes not. Like most teenagers, Elon hated going to bed at a reasonable hour. He read into the early morning – mostly science fiction, especially Isaac Asimov's *Foundation* series, featuring a bold visionary trying to stop the collapse of civilization.

Elon struggled to connect socially. His blunt and often impatient nature, coupled with his fascination for complex topics, made friendships difficult. She always knew he was a genius, his mother Maye later said. Other kids just saw a nerd. She remembered one of his cousins shouting, 'Look at the moon, it must be a million miles away!' Elon's response: 'No, it's like 239,000 miles, depending on the orbit.'

Classmates teased Elon and often assaulted him physically. Once, someone pushed him down a flight of stairs, leaving him unconscious in the hospital. The constant attacks made his school years a torment.

In the mid-1980s, when Elon was about ten, he decided to move in with his father – a choice he would come to deeply regret. His time with Errol was marked by psychological pressure and emotional abuse. 'It was terrible,' Musk told *Rolling Stone* magazine in November 2017. His father, he said, was a bad man – an evil man. 'Almost every crime you can possibly think of, he has done. Almost every evil thing you could possibly think of, he has done.'

After Elon was relentlessly bullied and beaten at his public high school, Errol decided to enrol him in a private one. Pretoria Boys High School, modelled on the British system, was known for its strict discipline. Teachers beat students with a cane, church attendance was mandatory, and uniforms were required.

At the new school, Elon earned top marks – except for two subjects: Afrikaans and religion. To his biographer Isaacson, he explained: 'I wasn't really going to put a lot of effort into things I thought were meaningless. I would rather be reading or playing video games.'

It was then that Elon discovered computers – starting with the Commodore VIC-20. For hours, he immersed himself in the world of programming, teaching himself to code. At 12, he created his first video game: *Blastar*. Elon sold it to a computer magazine for $500. His love of video games became a constant in his life – often with obsessive intensity.

At 17, Elon persuaded his mother to move to Canada. He handled all the paperwork and applications himself. In June 1989, as soon as the documents arrived, Musk set off. He began his studies at Queen's University in Kingston, Ontario, later transferring to the University of Pennsylvania in the U.S. There, he focused on physics and economics, already sketching out business ideas in the lecture halls. 'I was concerned that if I didn't study business, I would be forced to work for someone who did. My goal was to engineer products by having a feel for the physics and never have to work for a boss with a business degree.'

The Startup That Didn't Sleep

In 1995, Elon and his brother Kimbal Musk had an idea: a searchable online directory of businesses, like a digital version of the *Yellow Pages*. Elon had already enrolled at the elite Stanford University, but just two days in, he dropped out and founded his first company with his brother: Global Link Information Network.

The tiny office in Palo Alto had just enough space for two desks and a pair of futons. The Musk brothers slept there, showered at a local Young Men's Christian Association (YMCA), and lived mostly on fast food. A few months later, they rented an unfurnished apartment. According to people who were around at the time, Elon Musk regularly slept under his desk.

The breakthrough came after a visit to Navteq, a company with a database of land and road maps. Musk wrote a program that combined the maps with his business directory. Users could zoom in on sections of the map and move around with the cursor – much like we do today with Google Maps or other navigation apps. At the time, it was revolutionary. Musk patented the idea. In an era when most people barely used email, he was already far ahead of his time.

Musk's parents helped their sons. His father Errol contributed money, and his mother Maye flew in to help prepare for meetings. She often stayed up all night printing out presentation materials at a local copy shop.

Soon after, an offer came from Mohr Davidow Ventures: a $3 million investment. There were two conditions: The Musks' company was to be renamed Zip2; and, under pressure from the investors, Elon had to step down from the leadership role, making way for a more experienced manager.

Elon Musk became Chief Technology Officer – though he was convinced that wasn't enough. For a company to succeed, it needed him at the top. 'I never wanted to be CEO,' he tweeted in January 2022, looking back on his career. 'Running companies hurts my heart, but I don't see any other way to bring technology & design to fruition.'

And so, fate took its course. Musk drove the company forward day and night – weekends, holidays, and vacations be damned. He expected others to do the same. In the BBC documentary *The Elon Musk Show*, former Zip2 vice president Jim Ambras described what happened when someone didn't keep pace the way Musk expected.

'It was a Friday night. Elon was going around the office, looking to see who was sitting at their cubes,' Ambras said. 'There weren't many people sitting at their cubes. It was 9 p.m. – and his face turned red, he was really angry that the entire company wasn't there in the office at 9 o'clock at night.'

Musk's credo: a manager's job is not to be loved by their team. That applied to his brother, too. Unlike regular employees, Kimbal Musk didn't let Elon push him around. As biographer Isaacson tells it, their disagreements often escalated into physical fights. In front of their staff, the brothers wrestled on the floor.

Once, Kimbal bit Elon's hand and tore off a piece of skin, as he thought Elon might punch him. Elon had to visit the hospital and get a tetanus shot. He later commented unapologetically: 'Growing up in South Africa, fighting was normal. It was part of the culture.'

If Musk learned one thing from his first company, it was this: he wasn't made for the back seat. He disagreed with the new CEO's strategy. When the CEO decided to merge Zip2 with rival CitySearch, it led to a clash. Musk staged a rebellion that caused the deal to fall through. He demanded, unsuccessfully, to be reinstated as CEO. In 1999, the company was sold to computer manufacturer Compaq for $307 million. The deal made Musk a multimillionaire.

He wasn't satisfied. 'Great things will never happen with VCs or professional managers,' Musk told *Inc.* magazine. 'They have high drive, but they don't have the creativity or the insight.'

Musk soothed his bruised ego with the trappings of success. Not yet 30 years of age, he bought a 1,800-square-foot condo in Palo Alto and acquired a Formula 1 sports car from McLaren – Musk had CNN film the moment.

'Three years ago, I was showering at the YMCA and sleeping on the office floor, and now I have a million-dollar car,' he told a reporter. Musk insisted that becoming rich didn't have an impact on his core. 'Some could interpret purchasing this car as behavior characteristic of an imperialistic brat,' the 28-year-old speculated. 'My values may have changed but I'm not consciously aware of my values having changed.'

The Bank of Elon

After selling Zip2, Musk was ready for his next adventure. He wanted to revolutionize banking. In 1999, he founded X.com, then still an online banking startup. He invested $12 million – more than half his net worth. The idea: to create an online alternative to traditional banks. X was supposed to be faster, cheaper, and simpler. On one occasion, Musk stood in front of a group of skeptical investors, insisting that soon, no one would have to walk into a bank anymore to do banking. The investors laughed. Musk didn't.

One person who shared his vision was Peter Thiel. Now a Silicon Valley legend, Thiel too was just at the beginning of his investment career back then. Born in the German banking district in Frankfurt in 1967, Thiel moved frequently before his family settled in California in 1977. After high school, he studied philosophy and then law at Stanford University. He later worked for a federal judge in New York before switching industries and trading financial products at the Swiss bank Credit Suisse. In 1996, Thiel returned to California and went independent.

In 1998, he founded Fieldlink, which was soon renamed Confinity. His most successful product: the payment service PayPal. Thiel and Musk fought for dominance in the market, luring customers with ever more aggressive bonus offers. The companies burned through enormous sums of money. Musk later described it as 'a race to see who would run out of money the fastest.'

The pressure forced Musk and Thiel to the negotiating table. A merger between Confinity and X would create a giant. Musk was against the partnership, but the economic logic was so compelling that he even postponed his honeymoon in early 2000 to hash things out with Thiel. Soon after, they almost died on their way to more millions.

During a drive together to the investment firm Sequoia Capital, Thiel asked a question he would come to regret. Musk recounted the near-fatal episode in July 2012 with great enthusiasm during an interview with U.S. tech magazine *Pando Daily*.

'We were driving up Sand Hill Road and I didn't really know how to drive a McLaren,' Musk said. Thiel had asked him, 'So, what can this do?' Musk joked his response was probably number one on the list of famous last words: 'Watch this!'

He floored the accelerator. A moment later, the car's rear end swung out, hit an embankment, and the million-dollar beauty went airborne. The McLaren flew across the road, spinning like a discus, before slamming back down onto the pavement. Musk and Thiel were unharmed, but the McLaren – uninsured – was totaled. Thiel continued to the investor meeting in another car. Musk followed after dealing with the police and fire department.

Two months later, their companies merged. The name PayPal remained. This time, Musk made sure to place himself at the top – as CEO. Under his leadership, PayPal grew rapidly and quickly amassed millions of users. The service offered a simple way to send and receive money online and soon became the preferred payment method for online purchases – especially on the auction platform eBay.

In the wave of internet euphoria at the dawn of the twenty-first century, PayPal became Musk's first global hit. He worked obsessively. Everything had to move fast. In Ashlee Vance's 2015 biography of Musk, former PayPal employee Julie Ankenbrandt described the extraordinary intensity: 'We all worked 20 hours a day, and he worked 23 hours.'

When Musk did sleep, it was often at the office. He coded through the night and expected the same from his team. Those who couldn't keep up had to go.

As much as Musk impressed his new business partner Thiel with his drive and technical ability, he irritated him with his ego. Musk insisted that the merged company be called X – just like his own. Thiel couldn't understand why. PayPal already had a reputation in the community and, in his view, was the logical name for the new venture. The name stayed. So did Musk's ego.

After the merger, Thiel withdrew from daily operations and left it to his co-founder and CTO Max Levchin, a programmer from Ukraine, to stand up to Musk. The setup quickly led to a bitter power struggle.

Levchin grew increasingly alarmed as PayPal suffered mounting losses from fraudulent transactions and automated scams. Musk seemed largely uninterested. Instead, he spent an entire year debating the right programming language for the company.

By summer 2000, Levchin believed the company was in danger. He contacted Thiel with a bold plan: a coup against Musk. Thiel listened – and made a decision. While Musk was on his honeymoon in Australia, Thiel convinced the board to remove him as CEO. Then, Thiel seized the leadership for himself.

Musk called it a 'heinous crime,' then accepted the decision and asked to continue representing PayPal in public. No one could do it better, he argued. Thiel declined. In the span of three years, Musk had now been ousted from two of his own companies. One consolation: when PayPal went public in October 2002, Musk's shares were worth $180 million.

Alpha, Always

He may have been a social outcast in his youth, but as Elon Musk grew older, his air of genius gave way to an outsized self-confidence. Musk entered numerous relationships – sometimes several at once. His partners ranged from employees to Hollywood actresses. By the time he became a billionaire, Musk's ego seemed boundless. On September 11, 2024, he publicly offered to father a child with pop star Taylor Swift – in a single-line post on his social platform X.

The first woman with whom Musk had a child was Justine Wilson. He met the teacher while still a student at Queen's University in Canada. Musk was already an entrepreneur then, but not yet a millionaire. The couple shared an apartment with two roommates and an untrained dachshund. The following year, they married. The prenuptial agreement Musk presented led to a fierce argument.

Wilson prevailed, and in May 2002 she gave birth to their first son, Nevada. Ten weeks later, during a cousin's wedding, the boy died of sudden infant death syndrome. According to Wilson, Musk never wanted to talk about Nevada's death.

She later described their marriage as a rollercoaster. On the one hand, they lived in absolute luxury: dining at the most exclusive restaurants, sitting at best tables in the hottest clubs. One night they were placed next to *Titanic* director James Cameron; another, they partied with pop icon Paris Hilton or movie star Leonardo DiCaprio. The Musks were also on the guest list when Google founder Larry Page held his wedding on entrepreneur Richard

Branson's private Caribbean island — among guests like actor John Cusack and singer Bono.

At the same time, their marriage was a constant battle. In 2004, their twins Vivian and Griffin were born; in 2006, triplets Kai, Saxon, and Damian. Wilson stayed home; Musk flitted endlessly between his companies. Even at home, he couldn't switch off — always on edge, always wired. The couple fought like cats and dogs.

According to biographer Isaacson, Musk's own brother, who had once wrestled him to the floor, found the tension between the two so unpleasant that he avoided their company for a long time. Wilson accused her husband of ignoring her. He showed no interest in his children or his wife, she said — not even in the novels she wrote as an aspiring author. Musk claimed that his lack of empathy stemmed from his Asperger's syndrome — a form of autism that makes it difficult to read others' emotional cues.

Though Musk acknowledged his limitations, he preferred to flaunt them rather than apologize. Years later, during a guest appearance on *Saturday Night Live*, Musk quipped: 'To anyone I've offended, I just want to say: I reinvented electric cars and I'm sending people to Mars in a rocket ship. Did you also think I was gonna be a chill, normal dude?'

His relationship with his first great love was doomed to fail. When Wilson asked he spend more time with her and their children, Musk accused her of selfishness. Didn't she realize he was busy becoming a global entrepreneur? In September 2010, Wilson published an article in *Marie Claire* describing their marriage: 'As we danced at our wedding reception, Elon told me, "I am the alpha in this relation-ship." I shrugged it off [. . .] but as time went on, I learned that he was serious.'

Eventually, she understood that Elon would always be Elon. The drive to compete and dominate – the very thing that made him so successful in business – didn't shut off when he came home. 'Elon's judgment overruled mine, and he was constantly remarking on the ways he found me lacking. "I am your wife," I told him repeatedly, "not your employee." "If you were my employee," he said just as often, "I would fire you."' Musk filed for divorce in 2008.

Just weeks later, he met British actress Talulah Riley, 15 years his junior. Musk was in London giving a talk on commercial spaceflight – he had founded his space company SpaceX in 2002. The next evening, at a night-club, he met Riley. Musk seemed awkward, talked about rockets and stars, but the two hit it off. Two weeks later, they were engaged.

Like Justine Wilson, Talulah Riley later spoke candidly to the BBC about her relationship with Musk. She said she had been sure Elon was the one. But on her parents' advice, she didn't rush into marriage. The couple wed in 2010.

Musk's second marriage fared no better than his first. Occasionally, when Riley planned elaborate party trips, she could hold his attention for a while. Otherwise, Musk worked obsessively, barely acknowledging her presence. Even when she hosted guests, he stayed in his office, on the phone, giving orders – mentally everywhere but with her. In 2012, she moved out and filed for divorce.

Before the paperwork was even signed, the two were back in each other's arms. The divorce went through, but Musk and Riley continued living under the same roof. In 2013, they remarried. But whatever they hoped to hold on to elluded them. In the fall of 2015, Riley left Musk, and in 2016, she divorced him a second time.

Having one wife, five children, and multiple companies was not enough for Elon Musk to feel fulfilled. As early as 2012, he was pursuing actress Amber Heard, then in a relationship with actor Johnny Depp. In 2016, he courted her again; by 2017, they were dating. And if Musk's relationships were turbulent, his time with Heard could be described as a battlefield. As biographer Isaacson describes it, she and Musk would stay up all night fighting, and then he would not be able to get up until the afternoon. Fights escalated to the point where Heard locked herself in her hotel room, leading to security intervening to convince her she was safe. Later Heard described the episode as her being rather dramatic. The couple split in summer 2017, reunited briefly, and broke up again in early 2018.

The next woman in Musk's life was also an artist. Claire Boucher, born in 1988 in Vancouver, is an avant-garde musician and producer who gained recognition under the stage name Grimes. She later said she had offered Musk emotional support after his traumatic breakup with Heard – but their relationship grew. In 2020, their first son was born: X Æ A-Xii. Musk took a photo during the C-section and sent it to friends and family. 'It was Elon's Asperger's coming out in full,' Grimes told biographer Isaacson. 'He was just clueless about why I'd be upset.'

By then, Musk was 50 – but no less restless. In 2021, Grimes left him, unwilling to cope any longer with his relentless work schedule, absence, and emotional coldness. The breakup didn't last – and things remained complicated. In December 2021, they had a second child, a daughter named Exa Dark Sideræl, carried by a surrogate. Grimes later renamed her Y. Four weeks before her birth, Musk had become a father again – twice.

Shivon Zilis, a company executive, had given birth to twins: Strider and Azure. Musk apparently saw these children as gifts to humanity. While others worried about overpopulation, he believed men and women had a societal duty to reproduce, or else civilization would die out. Zilis, who deliberately lived without a partner, had wanted to become a mother. Musk offered to be the donor. His thinking: it's good for children to have intelligent parents. And, in his view, no one was more intelligent than he was.

Musk said nothing of this to Grimes – she learned about it from the media. By then, their third child was already on the way. The boy, named Techno Mechanicus, was born in June 2022, again via surrogate. In February 2024, Zilis gave birth to another daughter: Arcadia. A year later, when conservative influencer Ashley St. Clair announced she had conceived a child with Musk, whom they named Romulus, Zilis went public too. She revealed she'd had a fourth baby with the Tesla CEO: Seldon Lycurgus. He was Musk's 14th child.

Almost simultaneously, new Musk scandals surfaced. The U.S. outlet *Business Insider* reported that Musk had paid hush money to a SpaceX flight attendant. The reason: an incident in 2016. During a massage in a private cabin, Musk allegedly exposed his erect penis and asked her to 'do more' – offering to buy her a horse in return. The woman later hired a lawyer. The case was settled out of court. She received $250,000 in severance. The condition: she could not sue Musk or SpaceX, nor speak publicly about the incident.

When the accusations eventually became public, Musk dismissed them as 'politically motivated.' On May 18, 2022, he tweeted: 'In the past I voted Democrat, because they were (mostly) the kindness party. But they have become the party of division & hate, so I can no longer support them and will vote Republican. Now, watch their dirty tricks campaign against me unfold . . .' In another tweet, he added: 'Nothing will deter me from fighting for a good future and your right to free speech.'

Did Musk know what was coming? Two months later, the *Wall Street Journal* reported an affair. In December 2021, Musk had attended a party in Miami where he met Nicole Shanahan, wife of Google co-founder Sergey Brin. According to witnesses, the two disappeared for several hours. The *Wall Street Journal* and the *New York Times* reported that Shanahan had confessed a one-night stand with Musk to her husband. Both Shanahan and Musk publicly denied the incident. What is not in dispute: Shanahan and Brin separated two weeks after the party and later divorced.

Public attention focused not only on the alleged affair. Multiple guests also confirmed that Shanahan and Musk had taken ketamine – a sedative used in human and veterinary medicine, often abused as a party drug.

Altered States

Two hours into the interview, Joe Rogan reached for a joint. 'Oh, is that a joint or is it a cigar?' Musk asked. No, the podcast host clarified – it was a mix of tobacco and marijuana. Would Musk like a hit too? 'You probably can't, because shareholders, right?' Rogan said, while millions watched and listened. Musk hesitated for no more than two seconds. 'I mean, it's legal, right?' he asked. 'Totally legal,' Rogan replied. Musk took the joint and had a puff.

'How does that work – do people get upset at you if you do certain things?' Rogan asked. Musk took another drag, made a face, then shook his head.

The appearance, in September 2018, made headlines around the world. A corporate CEO smoking a joint on camera – when had that ever happened before? Some observers called for consequences, others thought it was just Musk's way of seeking attention. It wasn't the first time his relationship with drugs had raised questions. And it wouldn't be the last.

Reports about Musk's drug use began to pile up. Insiders, employees, and close friends described how Musk didn't just use soft drugs, but also experimented with LSD, cocaine, and ecstasy at private parties. These gatherings, often held under strict security, revealed a side of Musk that left both staff and business partners uneasy.

An incident at SpaceX in 2017 illustrated the growing concern in his inner circle. According to *The Wall Street Journal*, Musk arrived nearly an hour late to a meeting. When he finally took the stage before hundreds of employees, he appeared disoriented, slurred his words, jumped erratically between topics for 15 minutes, and repeatedly referred to the prototype of his rocket, the Big Falcon Rocket, as the 'Big Fucking Rocket.'

Several senior SpaceX executives suspected drugs were involved. While it was unclear whether Musk was under the influence that day, his behavior made people uncomfortable and sowed distrust inside the company. At the meeting in question, Gwynne Shotwell, SpaceX president, eventually decided to step in to take control. This was no way to do business.

The situation escalated further in 2018. After Musk's appearance on Joe Rogan's show, NASA grew wary. The agency – under contract with SpaceX for billions in astronaut transport to the International Space Station – viewed Musk's behavior as a liability. NASA required SpaceX to begin regular drug testing of employees. The agency even spent an additional $5 million on safety training at SpaceX to ensure the company complied with the rules.

Musk's drug use wasn't just a problem for government contracts. Similar concerns existed inside Tesla's boardroom. Insider accounts suggested that some executives viewed Musk's antics as increasingly erratic and unpredictable. Tesla shareholders questioned how Kimbal Musk, serving on the board, could realistically oversee his brother if the two had been seen together using drugs.

Chemical substances weren't the only issue. In his 800-page biography of Musk, Walter Isaacson uses the word 'addiction' only once – to describe Musk's insatiable appetite for video games. Like many kids, Musk spent hours immersed in virtual worlds. To begin with it was first-person shooters, later strategy and role-playing games like *Civilization* and *World of Warcraft*. Unlike most people, however, Musk couldn't moderate his gaming even at age 50.

In 2021, he reportedly developed an 'obsession with *Polytopia*,' a multiplayer strategy game on his iPhone. 'He doesn't have hobbies or ways to relax other than video games,' Musk's partner Grimes told biographer Isaacson. 'But he takes those so seriously that it gets very intense.' Once, when the two were playing together and Grimes didn't follow an agreed tactic, Musk didn't speak to her for the rest of the day.

During a visit to the Tesla Gigafactory in Germany, Musk reportedly delayed team meetings to keep playing. On the return flight, he spent the entire night glued to *Polytopia*. At the birthday party of his brother's wife, Musk sat hunched in a corner, glued to his phone, as the party went on around him.

Musk has never denied his love of games. In fact, he once claimed to be one of the best players in the world at *Polytopia*, good enough to beat its creator, Felix Ekenstam. The Tesla CEO also claimed to be one of the best players in *Diablo IV* and *Path of Exile 2*, a boast that would require thousands of hours of focused gameplay to back up. Seasoned gamers quickly challenged the claim, pointing to unexplainable mistakes. During a livestream of *Path of Exile 2*, viewers noted Musk's unfamiliarity with basic mechanics. This led to widespread speculation that Musk had outsourced the grinding process, paying others to level up his characters and acquire weapons and inventory far above his own skill level.

Whatever the case, Musk was unapologetic, both about the time he poured into games and the attention that they took away from his companies and family. 'I do play video games as my one recreational activity to quiet my mind,' he wrote in an X post in January 2025. 'Some days are real tough, so playing video games is my strange solace.'

In 2018, Musk's drug-related behavior seemed to reach a new level. He tweeted that he intended to take Tesla private at $420 per share – the number being a well-known code for marijuana use. The tweet triggered an investigation by the U.S. Securities and Exchange Commission, which resulted in a $40 million fine and Musk's temporary removal as Tesla chairman. Insiders later reported that Musk may have been under the influence of drugs during a follow-up interview with the *New York Times*. During the conversation, Musk broke into tears several times and spoke about the crushing burden of running Tesla.

In a 2023 tweet, Musk stated that ketamine was a better solution for depression than conventional anti-depressants. In an interview with former CNN anchor Don Lemon, posted on X in March 2024, he said he used the substance with a doctor's prescription to treat depressive episodes. 'There are times when I have sort of a . . . negative chemical state in my brain, like depression I guess, or depression that's not linked to any negative news, and ketamine is helpful for getting one out of the negative frame of mind.'

Musk also noted that he disclosed his prescribed ketamine use on X 'because I thought, maybe this is something that could help other people.'

When Lemon asked whether ketamine might 'get in the way' of his government contracts or affect his reputation on Wall Street, Musk denied that it would. 'From the standpoint of Wall Street, what matters is execution,' he said. 'Are you building value for investors?' He went on to highlight Tesla's strong valuation and sales. 'From an investor standpoint,' he said, 'if there is something I'm taking, I should keep taking it.'

Indeed, Musk's popularity seemed unaffected by his many scandals. Investors and shareholders appeared willing to overlook his extraordinary behavior – as long as his companies remained extraordinarily successful.

Elon's Final Frontier

In the early 2000s, Elon Musk set out to revolutionize the space industry. His goal: to lead humanity to the stars and make it a multiplanetary species, as he called it. To get closer to that dream, he moved to Los Angeles. His then-wife Justine suspected he was drawn by the glamour. But for Musk, it was a strategic move. Here, he had access to top talent and the greatest aerospace resources in the country.

At first, he envisioned a philanthropic mission called

Mars Oasis, which would send a small greenhouse to the Red Planet to grow plants. Musk traveled to Russia in hopes of buying decommissioned intercontinental missiles at a bargain price.

The negotiations didn't go as planned. According to his adviser Jim Cantrell in a later documentary, Musk showed up in his usual casual clothes – something the Russians saw as disrespectful. They also didn't believe the young man could actually pay the millions they were asking. 'They looked at us like we weren't credible people. They thought we were full of shit,' Cantrell said. 'One of their chief designers spit on me and Elon.'

Musk turned the insult into fuel. On the flight home, he dove into technical specs on his laptop, built spreadsheets of material costs, and reached a conclusion: 'I think we can build this rocket ourselves.'

In May 2002, Musk founded Space Exploration Technologies – SpaceX. He made himself CEO and chief engineer. Later, he recalled how friends and advisors warned him against investing in space. They thought it was the fastest way to torch his PayPal fortune.

And indeed, the early years were brutal. The first three test launches of the Falcon 1 failed, nearly sinking Musk and his company. 'The likeliest outcome is that I will lose all my money,' he told his longtime friend and backer Adeo Ressi. 'But what's the alternative? That there be no progress in space exploration? We've got to give this a shot.'

On the fourth attempt, in September 2008, the Falcon 1 reached orbit. It was a turning point. Soon after, Musk landed a $1.6 billion contract with NASA to deliver cargo to the International Space Station. He had made it – all in, all out, and all the way.

Over the next few years, SpaceX focused on developing a new Falcon rocket and the Dragon spacecraft. Musk stayed closely involved in the engineering and relentlessly pushed his team forward. On December 8, 2010, the Falcon 9 launched with Dragon. On May 25, 2012, it became the first private spacecraft to dock with the ISS. Then, on December 21, 2015 – after several failed attempts – SpaceX successfully landed the first stage of a Falcon 9 back on Earth. It was a breakthrough that changed the space industry. SpaceX could now reuse its rockets. Costs dropped dramatically.

Meanwhile, Musk conceived Starlink, a satellite network designed to provide global internet access. He saw it as a way to advance technology – and raise money for a Mars mission. In May 2019, SpaceX launched the first 60 Starlink satellites. By January 2020, there were over 240 in orbit. The goal: to deliver internet service to even the most remote regions on Earth.

In February 2022, after Russia invaded Ukraine, Starlink took on a sudden political and military role. The Kremlin targeted Ukraine's digital infrastructure. SpaceX sent thousands of Starlink terminals to the country, restoring connectivity even as ground networks were destroyed.

Afterwards, he seemed dazed by his influence. 'How am I in this war?' Musk asked his biographer Isaacson. 'Starlink was not meant to be involved in wars. It was so people can watch Netflix and chill and get online for school and do good peaceful things, not drone-strikes.'

And yet, this was his new reality. Musk was worth over $200 billion, controlled multiple multi-billion-dollar companies, and was helping shape the fate of nations. But that didn't shield him from personal failings.

In June 2024, the *Wall Street Journal* reported that Musk had a 'romantic relationship' with a former SpaceX intern. He was also accused of initiating a sexual relationship with a subordinate and of pressuring other SpaceX employees. One woman said Musk repeatedly asked her to 'have his babies.' Eight female former employees filed a lawsuit against Musk for sexual harassment.

The Electric Empire

The story of Tesla didn't begin with Elon Musk. Its origins trace back to Martin Eberhard. The engineer from Silicon Valley simply wanted to buy an electric car. When he couldn't find one, he decided to build it himself. His partner was a friend: software engineer Marc Tarpenning. In 2003, they founded a company and named it after inventor Nikola Tesla. They had a big idea – but only small funds. Through an intermediary, they came into contact with Elon Musk. Fresh from the founding of SpaceX, a year earlier, Musk still had both cash and momentum. After just one conversation, he decided to invest in Tesla.

The partnership was rocky from the start. On paper, Eberhard was CEO, but Musk, as an investor, intervened in almost every detail. He had never designed a car before, so he read every book he could find on the subject and studied the design of iconic brands. Then he refused to be told he was wrong – least of all by the man whose salary he was paying. Door sizes, dashboard layouts – Musk believed he had the best solution for everything. Eberhard was frustrated. His biggest concern: Musk's endless revisions and additions threatened to drive the price so high that most buyers couldn't afford it.

Musk's priorities were different. The first Tesla didn't need to be affordable – it needed to be perfect. He pushed Eberhard to hire more people. By spring 2006, Tesla had 70 employees – but no product. The management had to inform the board that the company was running out of money. Musk erupted in a tirade, then raised a fresh $40 million in a new financing round. The press release announcing the news drove a wedge between Eberhard and Musk.

'Tesla Motors was founded in June 2003 by Martin Eberhard and Marc Tarpenning,' it read. The management thanked Musk for his trust – as an investor and lead board member. Musk saw it as an insult. From that point on, he gave interviews about Tesla without clearing them with anyone. In July 2006, when the first prototype of the Tesla Roadster was set to be unveiled, Musk scrapped the communications chief's PR plans and ordered Eberhard to fire her. Then he took over – down to the guest list, the menu, even the napkin design.

The launch was a success, but Musk felt slighted. Both Eberhard and Musk gave speeches at the Roadster unveiling. The press loved it – but focused almost entirely on Eberhard: 'He set out to build a sleek, battery-powered performance machine,' *Wired* wrote. Musk was mentioned only as an investor. In an internal memo, he complained bitterly. The coverage was 'outrageous' and 'incredibly insulting' he said. His influence on the Roadster ranged 'from the headlights to the styling to the door sill to the trunk, and my strong interest in electric transport predates Tesla by a decade.' He announced he would speak to 'every major publication within reason' to set the record straight.

The next day, the *New York Times* ran a story about Tesla – without even mentioning Musk. The only photo showed Eberhard and Tarpenning. Musk fumed that the article was an incredible humiliation and warned that 'if anything like this' happened again, he would not only fire the communications chief, but also the external PR agency.

In November 2007, Tesla announced Eberhard's departure as CEO. In an interview with *Business Insider*, Musk called Eberhard 'literally the worst person I have ever worked with' and added that Eberhard had come 'damn close to killing Tesla through a combination of bad management decisions, driving out talented people, bad engineering, major blunders in supply chain & an elaborate deception about the true cost & schedule of the Tesla Roadster.'

The wounds never fully healed. In June 2009, Eberhard sued Musk for defamation, slander, and breach of contract. The dispute ended in a settlement. But even in 2022, Musk referred to Eberhard in a tweet as 'a compelling liar.' Eberhard later said in interviews that Musk had exaggerated his role in Tesla's early years. 'So, you know, the idea of him, like, sitting around working on the car or something is simply not true,' Eberhard said. 'He was not there.'

That changed in 2008. Musk became CEO – and was immediately confronted with the threat of bankruptcy. In a new funding round, he raised only $20 million. He had no money left of his own. Musk had to ask everyone he knew for help. Google co-founder Sergey Brin gave him $500,000. Many Tesla employees pitched in with personal loans. The parents of his wife Talulah even offered to mortgage their home for him. In interviews, she recalled: 'Every day Elon would come home and say, "That's it." He was

being ridiculed by the press. He would have these night ter-
rors in the middle of the night. We'd be fast asleep and
suddenly he would be screaming in his sleep, like he was
trying to climb up and escape something.' According to his
wife, Musk even offered to let her leave, she said – to spare
her from this phase of his life. She stayed, but feared for the
worst. 'I was worried he's gonna have a heart attack, all the
available resources had to be plowed to Tesla and SpaceX.'

In the end, it was the U.S. government that bailed Musk
out. In June 2009, with Tesla nearly out of cash and private
funding drying up, the U.S. Department of Energy stepped
in with a $465 million low-interest loan. The financial crisis
had scared off most investors, and Tesla was on the brink –
barely making payroll, with suppliers growing nervous. The
government lifeline didn't just fund operations; it helped
him survive. The public money gave Tesla a chance to
stabilize, regroup, and push Musk's vision forward. Without that loan, Tesla might have been remembered as just
another failed startup with big ideas and no runway.

After years of torment, Musk finally had his breakthrough in 2012. Tesla launched the Model S, a sleek,
high-end electric car that drew global attention. It won
a slew of awards and proved that electric vehicles could
be stylish and high-performing – not just eco-friendly.
Within a year, the Model S became a bestseller in the U.S.,
outpacing every other luxury sedan – from BMW and
Mercedes to the top domestic names. Musk had become

a serious player in the automotive industry – a world that had been unaccustomed to new contenders for decades.

The rise was meteoric. In 2013, Tesla brought in $2 billion in revenue – five times more than the year before. In 2014, it topped $3 billion. In 2015, $4 billion. Tesla was still a minor player among automakers – but no one else was growing this fast. Driving a Tesla became a statement. And Musk began to claim not just that he was building the best cars in the world – but that, by doing so, he was saving humanity.

In 2017, Tesla introduced the Model 3 – the first electric car aimed at the mass market. Demand was overwhelming, but Tesla hit another crisis. The Model 3's production was so complex that the company fell months behind schedule. Musk slept at the factory, on the floor of the production line, working around the clock until the problems were fixed. In 2018, Tesla passed $20 billion in revenue. By 2022, when our research into the Tesla Files began, this number had grown to more than $80 billion – with $12.6 billion in profit.

At the end of 2023, Musk gave an interview to the *New York Times*. 'Tesla currently sells twice as much in terms of electric vehicles as the rest of electric carmakers in the United States combined,' he said. 'Tesla has done more to help the environment than all other companies combined. So it would be fair to say that therefore, as a leader of the company, I've done more for the environment than any single human on earth.'

In 2024 – a year marked by falling revenues, shrinking market share, and cost-cutting – Elon Musk stood to receive an astonishing payday. Six years earlier, the board had approved a massive pay package to keep Musk focused on Tesla. Provided he met several ambitious targets, he was eligible to receive stock options worth up to $56 billion. By 2022, Musk had delivered – and was due to be awarded more money than any CEO in history. That's when the courts stepped in.

One investor sued, arguing that the goals for Musk hadn't been transparently communicated and that the board was too beholden to the CEO. The Tesla board included Musk's brother and several men he had made extraordinarily rich. There were also rumors of excessive parties and possible shared skeletons in Tesla's cellar. Was this a case of see nothing, hear nothing, say nothing?

In January 2024, a court struck down the pay package, calling the board's process 'deeply flawed.' Despite the ruling, Tesla's board put a revised version to a new vote – this time worth $45 billion. Major shareholders like Norway's sovereign wealth fund and California's CalSTRS pension fund announced they would vote no. Leading proxy advisors Glass Lewis and Institutional Shareholder Services also criticized the package as excessive.

In June, the vote went Musk's way. He thanked shareholders with a tweet: 'I love you guys!' Yet when the package came back before the same court, the judge didn't see how anything had changed. In December 2024, Delaware Chancellor Kathaleen McCormick reaffirmed her earlier decision to void the pay package, citing Musk's significant influence over the board and the lack of independent oversight.

Tesla appealed. The final word now lies with the Delaware Supreme Court. No timeline has been set.

Beyond the Brain

Musk's work at companies like Tesla, SpaceX, and Twitter could easily fill several full-time jobs. But no matter how many challenges he faces each day, Musk has never shied away from taking up new ones.

In 2016, frustrated by the traffic in Los Angeles, he founded the Boring Company and began planning tunnel systems to move vehicles underground at high speed. His vision was to create a transportation system that would shift traffic off the roads and into a subterranean space. As with so many of his ventures, Musk framed the project as a contribution to humanity's progress. On the other hand, in April 2024, the National Council for Occupational Safety and Health cited the Boring Company for major workplace safety violations.

Neuralink, also founded in 2016, may be Musk's boldest and most futuristic venture yet. Its mission: to link the human brain directly to computers. Musk's aim is to help people with neurological conditions restore or enhance brain function. In January 2024, Neuralink announced its first successful operation. Following the procedure, Noland Arbaugh, a 30-year-old American paralyzed from the shoulders down after an accident, was able to control a computer cursor with his mind. He used his new ability to play chess and learn foreign languages. Not long after, most of the 64 hair-thin threads implanted in his brain detached, requiring further surgery. In August 2024, Neuralink carried out a second implantation on another subject.

But Musk doesn't just want to help the sick. The long-term goal of Neuralink is to merge human intelligence with artificial intelligence. Musk argues that this kind of symbiosis is 'essential to the survival of our species.'

According to Musk, the future of humanity also depends on OpenAI. He co-founded the organization in 2015, alongside Sam Altman, as a nonprofit with the goal of developing artificial intelligence for the benefit of humanity. Musk was concerned about the dangers of a coming 'superintelligence' that could turn against human beings. Soon after OpenAI was established, Musk began to question the organization's direction and long-term goals. He left the company in 2018 and warned that artificial intelligence could become humanity's greatest threat. OpenAI's decision to commercialize AI, he said, was dangerous.

In March 2024, OpenAI released emails showing that Musk had tried to merge the company with Tesla. Lawsuits and countersuits followed. Musk has since launched his own artificial intelligence company: xAI. In May 2024, he announced that he had raised $6 billion from investors for the venture.

Go Fuck Yourself

In April 2022, Elon Musk stunned the world by announcing his intention to acquire Twitter. The news triggered a global wave of mixed reactions. Some users and investors cheered the prospect that Musk might revolutionize the social media platform. Others feared that his erratic style and often controversial tweets would threaten Twitter's integrity and stability. To some, he was the savior of free speech; to others, a poison for public discourse.

Musk's relationship with Twitter had always been turbulent. In 2018, a Thai youth soccer team was trapped in a flooded cave after heavy rains. Musk offered to send a mini-submarine to assist in the rescue – an idea dismissed by the man in charge. Vernon Unsworth, one of the most experienced divers in the world and coordinator of the Thai rescue mission, declined Musk's offer as a 'PR stunt' with no real chance of success. Musk responded by calling Unsworth 'pedo guy' on Twitter, implying he was a pedophile. In the end, Unsworth and his team managed to save the children after more than two weeks underground – without Musk's help.

The British diver later sued Musk for defamation in the U.S., but lost. Musk told the federal court in Los Angeles the phrase 'pedo guy' was common in South Africa, where he grew up. Speaking to reporters outside the courtroom, he said: 'My faith in humanity is restored.'

Unsworth's lawyer criticized the outcome as harmful for society, saying: 'This verdict sends a signal, and one signal only – that you can make any accusation you want to, as vile as it may be and as untrue as it may be, and somebody can get away with it.'

In 2020, Musk tweeted that Tesla's stock price was 'too high,' triggering an immediate market drop. During the Covid-19 pandemic, he downplayed the virus and opposed lockdowns. In 2021, he engaged in a public spat with U.S. senator Bernie Sanders, tweeting him to ask whether he was still alive.

In April 2022, Musk offered $44 billion to buy Twitter. He already owned 9 percent of the company's shares, making him its largest stakeholder. In July, he abruptly tried to back out of the deal, claiming Twitter had breached its obligations by failing to act on fake accounts. Twitter sued to force the acquisition. Just before trial, Musk agreed to complete the purchase under the original terms.

On October 27, 2022, Musk took control of Twitter and became CEO. He immediately fired several top executives and announced mass layoffs affecting about half the staff. The result: major disruptions to the company's operations. One of his most controversial decisions involved a new verification policy. Users could now simply buy the blue checkmark that once signaled account authenticity, leading to a flood of fake accounts. Many users suddenly found themselves on a platform overrun by deception.

Despite the chaos, Musk demanded absolute loyalty and commitment from his employees. On November 16, 2022, he sent an email with the subject line 'A Fork in the Road.' It read: 'Going forward, to build a breakthrough Twitter 2.0 and succeed in an increasingly competitive world, we will need to be extremely hardcore. This will mean working long hours at high intensity. Only exceptional performance will constitute a passing grade.'

Employees who were certain that they wanted to become a part of X were asked to click on a link in the email. Musk: 'Anyone who has not done so by 5pm ET tomorrow (Thursday) will receive three months of severance.' Employees fled in droves – so many that the company later tried to rehire some of them.

In December 2022, Musk launched a Twitter poll asking whether he should step down as CEO of the social media company. The majority voted yes. Musk made no move to follow their will. Throughout 2023, he experimented with new technologies to make Twitter more profitable. Finances remained tight.

Layoffs and an exodus of advertisers had left gaping holes in Twitter's accounts. Musk sought new revenue streams by introducing subscription models and promoting content monetization. Still, financial troubles were ever-present. Musk's vision of turning Twitter into an 'everything app' that would encompass all aspects of digital life began to seem illusory. In June, he handed the CEO title to former NBCUniversal executive Linda Yaccarino. But Musk's behavior – on and off the platform – left no doubt about who remained in charge. With over 150 million followers, Musk was the platform's most prominent voice by the end of 2023 – his posts reached more people than those of former U.S. president Barack Obama, footballer Cristiano Ronaldo, or pop stars Justin Bieber and Taylor Swift.

Often, Musk seemed addicted not only to fame and attention but also to outrage. In November 2023, a user on the now-renamed platform X posted that Jewish communities were pushing 'the exact kind of dialectical hatred against whites that they claim to want people to stop using against them.' Musk replied: 'You have said the actual truth.'

The backlash was immediate. Musk's response was widely seen as promoting an antisemitic conspiracy theory. Jewish organizations and civil rights groups condemned the remark. Prominent Jewish leaders and activists demanded an apology. The Anti-Defamation League and others voiced deep concern about the spread of antisemitic rhetoric by such a powerful figure. Companies like Apple and Disney pulled their advertising from X.

At the end of November, the *New York Times* invited Musk to its DealBook Summit. Reporter Andrew Ross Sorkin sat in a suit and tie on stage. Musk looked as though he'd just walked off a *Top Gun* set. He wore a leather bomber jacket with a fur-lined collar, a black shirt, black jeans, and boots.

Sorkin introduced Musk as a superstar. 'He's the richest person in the world. He may very well be the most consequential individual in the world right now,' the reporter said. 'He runs the most innovative companies in the world.'

Musk was disruptive, moving at breakneck speed, Sorkin told his audience. 'But he's facing some controversy in the process.'

Musk grinned. He had just returned from a trip to Israel, where he had met with the country's prime minister and president. Sorkin said the plan was to talk about all of it. Some had criticized the *New York Times* for giving Musk a platform, considering his recent remarks about Jews. 'I know you have an issue with journalists oftentimes, but I said it's our role to have conversations and to inquire and sometimes even interrogate ideas and I'm hoping we can do that.'

He got straight to the point. Sorkin asked Musk to walk through what had happened in the past ten days. What had led to the post about Jews and what came after. The firestorm of criticism, his subsequent trip to Israel.

Musk denied a connection. 'Well, the trip to Israel is independent of . . . It wasn't some apology tour. I want to be clear. [. . .] It wasn't in response to that at all,' Musk said. Then he added: 'And I have no problem being hated, by the way.' He looked at Sorkin, then out into the crowd. 'Hate away.'

Sorkin hesitated. Before he could ask the next question, Musk added, 'I think it's a real weakness to want to be liked, a real weakness. I do not have that.'

'OK,' Sorkin replied. 'Let me ask you this, then. There's a difference between saying, I don't care if anyone likes me or they hate me. But given your power and given what you have amassed and the importance you have . . .' He paused. 'I would think you want to be trusted. I would think maybe you don't need to be liked or hated, but trusted matters. If X is going to become a financial platform where people are going to put their money, where the government's going to give you money for rockets, where people are going to get into the cars, they need to ultimately decide that you are . . . They don't have to say that they love you, but that you are ultimately a decent and good human being.'

Musk was fast to answer. 'Yes. I mean, I think I am, but I'm certainly not going to do some sort of tap dance to prove to people that I am.' He rattled off his companies, his achievements. SpaceX built the best satellites. Tesla made the best cars. Whether people trusted him was irrelevant. His products spoke for themselves. He would not pander. Then Musk started to talk about his trip to Israel.

Sorkin listened, then asked, 'What was that trip like? And obviously you know that there's a public perception – and you're clarifying this now – but there's a public perception that that was part of an apology tour, if you will. This had been said online. There was all of the criticism, there was advertisers leaving. We talked to [Disney CEO] Bob Iger today . . .'

Musk cut him off. Then came a moment this stage had never seen before.

Musk: I hope they stop.

Sorkin: You . . . hope?

Musk: Don't advertise.

Sorkin: You don't want them to advertise?

Musk: No!

Sorkin: What do you mean?

Musk: If somebody's going to try to blackmail me with advertising? Blackmail me with money? Go fuck yourself!

Sorkin: But . . .

Musk: Go fuck yourself. Is that clear? I hope it is.

CHAPTER 3

A Nerve-Racking Investigation

Sönke's gift from Norway multiplies our database a hundred-fold. Until now, we had only combed through information about employees. Suddenly, the stream of data expands to include customers, business partners, and a wide array of individuals and companies with ties to Tesla. The hard drive from Drammen contains original documents from law firms, banks, and government agencies, presentations on technical details, summaries of autopilot failures and internal memos on a dozen subjects.

It's 5:35 p.m. when Sönke messages the investigative team: 'Hey all, the 19 gigabytes are uploaded. Merry Christmas!'

Although most of us are already on holiday and it's the day before Christmas Eve, our minds stay on the job. Laptops flip open. Between trimming trees and wrapping presents, we exchange notes on first discoveries, sketch plans for analyzing the files, and discuss safety protocols. The spirit of the season is quickly replaced by adrenaline – especially when the source, evidently feeling more confident after the meeting in Norway, sends a second batch of files the very next day. For many of us, gift-giving is already underway when Sönke drops another line in the team chat:

'He sent more files. Enjoy dinner, everyone!'

We're fully aware of what's at stake. If the Tesla Files are genuine – if what the whistleblower claims is true – we now hold technical and financial data that could be worth millions, even billions, to Musk's rivals. There's also personal information on tens of thousands of individuals. The kind of information that cybercriminals would love to get their hands on. If we question Tesla's competence protecting its customers and employees, we cannot afford to be careless ourselves.

That's why, from the very beginning, we take every precaution to keep the data secure. The raw files remain to this day on a computer purchased solely for this purpose – offline, never connected to the internet, never removed from the *Handelsblatt* building. We store the laptop in a safe with access restricted to fewer than five people.

At the same time, we need a secure working environment – one where every *Handelsblatt* reporter involved can do their work, whether they live in Berlin, Düsseldorf, Tokyo, New York, or San Francisco. So we also upload the data to *Handelsblatt*'s internal IT system. But, unlike Tesla, we implement strict security protocols. Access is restricted to just our team. No laptop or smartphone is allowed to connect automatically to the *Handelsblatt* cloud. Everyone must enable multi-factor authentication. Even if a device is stolen – even if someone knows the password – no one can access the Tesla Files without the second verification factor.

Fear of the Zip Bomb

Lukasz gave us a dataset he had compiled gradually over time. It arrived in the form of several zip files – a kind of digital container. Zip archives have the advantage of compressing their contents to save storage space. But anyone wanting to work with them has to extract the contents first. We scan the main file for viruses and begin to unpack, expecting individual files. However, we discover that the very first zip file we open contains more zip files, which in turn contain more zip files – and so on.

For a moment, we worry that *Handelsblatt* might have fallen victim to what's known as a zip bomb – a malicious file designed to trap its recipient in an endless loop of unpacking, consuming massive amounts of memory and crashing the targeted system. Zip bombs are a popular tool among hackers who want to paralyze networks.

In our case, the structure turns out to be less an attack than an attempt at order. Lukasz, the whistleblower from

Norway, simply tried to archive his growing trove of material in a somewhat organized fashion. In doing so, he created duplicates that now cause repeated error messages during extraction. It takes hours of decompression before we can grasp the true scale of the data. We upload the files to the cloud and store the raw data on the research laptop. Then we delete all digital copies.

As much as we welcome the glimpse into Tesla's inner workings, the dataset is small by comparison. In previous investigations, we've handled far larger volumes of confidential material. During the Wirecard fraud scandal, we collected over one and a half terabytes of data. In the case of the real estate group Adler, a few hundred gigabytes less – but still more than ten times what Lukasz has provided.

And yet, working with the Tesla Files proves unusually demanding. Even after extraction, the data feels completely random. The only common thread is that it allegedly comes from the heart of Musk's empire. Beyond that, there's no coherence – neither in file types, naming conventions, nor content. Many files aren't original documents, but screenshots, phone photos, or video recordings of other documents. They date back to a time when Lukasz was still fearful of being caught in the act. Some files are video clips several minutes long, showing spreadsheets with hundreds of thousands of rows.

A later review will show that the archive contains more than 23,000 files. Among them are over 10,000 image files,

1,700 text documents in PDF or Word format, more than 1,000 Excel spreadsheets, and 200 PowerPoint presentations. There are also hundreds of video and audio files. In addition, the dataset includes more than 1,700 archived web pages – allegedly from inside Tesla's own IT systems.

The documents include invoices, bank statements, contracts, engineering schematics, customer feedback summaries, and spreadsheets calculating the cost of car parts. Some files contain internal correspondence between dozens of Tesla employees over multiple years, all focused on a single technical issue. Some are sorted by customer, others by vehicle identification number. Many exist only as scans and, like Lukasz's screenshots and videos, must be processed with text-recognition software before we can work with them at all. Files that cover similar content are scattered across entirely different directories.

In short: the data is in chaos.

We quickly realize that keyword searches alone won't get us far. We agree that we'll have to open every single file at least once. The risk of missing something is simply too high. In one of the many team meetings we hold at the time, we abandon the idea of setting a publication date. We have to admit to ourselves that even just sorting through and structuring this material will take weeks.

Tesla's Holy Grail

Datasets like the Tesla Files make a company partially transparent. But anyone hoping to make sense of it must first acquire an entirely new body of knowledge. Corporations often develop their own internal language – complete with unique vocabulary and acronyms. So, right from the start, we create a glossary to collect and define Tesla-specific terms. We learn that PRP stands for Product Privacy, VQ for Vehicle Quality, and BA for Business Analytics. And N1? That's Number One – the internal shorthand for the boss: Elon Musk.

Some documents are nearly indecipherable at first – but the more we read, the more patterns emerge. Gradually, we begin to see how files are connected. Lukasz often downloaded a central document from Tesla's internal systems, and elsewhere we find its matching attachments. He followed keyword after keyword, chasing leads, building his archive step by step.

Some files we first dismissed as unremarkable turn out to be essential. We dig them back up once we realize just how central they are. Eventually, we crack the logic behind the archive's naming conventions: the numeric codes in hundreds of Excel spreadsheets map directly on to specific technical issues – and within the sheets we find customer complaints that match them.

We find the names of thousands of Tesla drivers whose vehicles allegedly braked or accelerated without cause. A presentation dated June 17, 2021, addresses issues with brakes that squeal, grind, or jolt. Its title: 'Worst things we are doing to our customers.' A 40-page PowerPoint by an electrical engineer in Palo Alto, dated May 2018, outlines the 'Battery Monitoring Architecture' for the Tesla Model 3. It compares the cost of reliability with the cost of warranty claims – and openly calculates where there are opportunities to compromise on quality.

We come across documents with specifications and measurements for the 'Roadrunner' battery module. A June 2020 presentation titled 'Visual Assembly & Process Flow' describes the production process in 40 illustrated steps. As we dig deeper, an uneasy feeling creeps in.

The Roadrunner project is considered Tesla's holy grail: a next-generation battery cell, with higher energy density and significantly lower production costs. Competitors would pay a fortune for these files. Again and again, we ask: how could a low-level employee walk out the door with all of this – on a USB stick?

As we search for information about Tesla's German Gigafactory in Grünheide, we find confidential details – including specifics on the paint line. One document from early 2020 records Elon Musk's directive: 'We need a color in Berlin that is rich, deep, radiant and really pops.' The goal set by the boss is 'Rolls-Royce-quality paint at Tesla prices.'

We uncover even more records containing sensitive personal data. Social security numbers, government IDs, private addresses, salaries, and stock allocations. Customer data is abundant. Spreadsheets list more than 100,000 names – including private email addresses and phone numbers, among them those of actress Talulah Riley, Musk's ex-wife, and Apple co-founder Steve Wozniak.

Over the holidays, our IT system shows a flurry of activity. Colored dots with our initials indicate who is looking at which file. Even late at night, we sit on our couches, laptops open, immersed in a rare and revealing glimpse into the Tesla machine. Many files are labeled 'Proprietary and Confidential Business Information,' often followed by: 'Attorney-Client / Work Product Privilege.' Lines of numbers and code flicker across our screens – more captivating than anything on Netflix or the books stacked on our nightstands.

On Threema, the encrypted messenger app, we share the 'nuggets' we've found with each other – the most exciting discoveries from the data. Our exchange with whistleblower Lukasz Krupski also remains active. On December 24 alone, he sends several dozen messages. One of them carries a surprise: Lukasz has found another hard drive – this one with 49 gigabytes of data.

Exposed by Default

It's time to speak again with our editor-in-chief, Sebastian Matthes. He remains skeptical; many companies have been plunged into chaos by data leaks. But this one is different. The information in question, some of it highly sensitive, apparently wasn't stolen by hackers. It was – so the whistleblower claims – simply accessible and copyable inside Tesla. It sounds too simple to be true. Matthes gives us a clear directive: gather more evidence that the data is authentic – and that the whistleblower is telling the truth about its origin and Tesla's lax data protection.

Now, we face a dilemma. We can't just call the people named in the files and ask whether the information about them is correct. It would be the most direct route, but if Tesla finds out about our investigation at this early stage, the possible repercussions are enormous. The company could take legal action or retaliate against the whistleblower. That's a risk we can't take – not at this stage.

So, we avoid reaching out to current or former employees. Many are deeply loyal to Musk and Tesla. It would only be a matter of time before someone passes word up the chain. Nor do we contact customers. We can imagine how they'd react if a journalist called them out of the blue and asked them private questions that can mean only one thing: a data leak. We need another approach.

Our first move is to cross-reference the Tesla Files with public data – on a massive scale. We check job titles on LinkedIn, compare employment dates, and make sure the records match. We also spend time verifying the internal logic of the dataset, even though it's messy. In the metadata of a PowerPoint file, for instance, we find the author. Does it make sense that this employee worked on that topic? Is he part of the right team? Does his name appear in other files about the same issue? Are there internal conversations where he's mentioned? We gather as many corroborative examples as we can.

It's the season of AI hype. Just weeks earlier, OpenAI released ChatGPT to the public. Suddenly, everyone is experimenting with artificial intelligence. The internet is flooded with breathless takes about its power, its dangers, its world-changing potential. We discuss it often. Could AI fabricate a complex, massive dataset like the Tesla Files? Theoretically, yes. If we can find job titles and teams on LinkedIn, so can a well-trained AI bot. It could plausibly simulate who works on what. Our cross-referencing helps, but it's only circumstantial. Not proof.

Lukasz has told us that most of the material comes from Tesla's project management system Jira, developed by Australian software company Atlassian. It's widely used – according to Atlassian – by more than 100,000 companies around the globe. Tesla competitors like Audi and BMW use it, as do Deutsche Bank and Twitter. By chance, we learn during a conversation in the newsroom kitchen that *Handelsblatt* uses Jira too – not in editorial, but in other departments. It's a reminder that office small talk can be valuable. A new path opens: maybe our own colleagues can help us make sense of the inner workings of Jira.

Originally built for software developers, Jira allows users to create 'tickets' – tasks that can be assigned to others, tracked, and documented collaboratively. It's meant to streamline work across teams and offices. According to our files, Tesla uses Jira far beyond engineering. Tickets cover everything from battery modules to insurance claims to customer-service quality. Many carry tags like 'Business Critical' or 'HR-Confidential.'

A ticket from New York, dated April 3, 2021, asks whether HR reports could include fields like 'Ethnicity,' 'Hispanic or Latino?' and 'Military Status.' In San Francisco, a technician reports a furnace leaking a 'pudding-like dross' on November 4 of the same year. Seven weeks later, a staffer in Barcelona asks for help from the security team. A colleague had emailed to report that the service center in Norway had been left 'unarmed since Sunday' because the alarm system wasn't working. Eventually, someone from the Gigafactory in Germany steps in remotely.

There's an old trick in business journalism: compare what a company does to what it says it does. Now that we've combed through the Tesla Files, we check what the automaker publicly claims about data handling. We can't help but smirk.

All data collected, created, or stored by Tesla 'must be kept confidential,' the company states on page 9 of its Code of Ethics – because that information 'contributes to Tesla's business success.' Employees are reminded that 'the outside world is intensely interested – and in some cases borderline obsessed – with what we do at Tesla.'

One internal policy outlines how to handle sensitive data: personal information such as Social Security numbers or passport IDs may only be shared with a password and with a supervisor's permission. Access is to be granted solely on a strict 'need to know' basis. According to the same document, Jira tickets – due to their 'sensitivity and risk to Tesla if mishandled' – are ranked among the highest security levels. Mishandling can result 'in disciplinary action, up to and including termination of employment.'

By that logic, Jira is the right choice. It's designed to give users highly specific access rights. Atlassian promotes this as a key feature that allows users to 'control what users within those applications can see and do.' But how exactly Tesla configured those settings – and how we can verify this – is unclear.

We turn to our colleagues at *Handelsblatt* who manage Jira internally. Normally, editorial and business operations are strictly separated to protect journalistic independence.

But in this case, the business side might hold the key. Our colleagues explain how Jira permissions are managed – and even offer to contact Atlassian support for more technical answers. We don't mention Tesla.

Within a week, we get a reply. Atlassian recommends using 'security levels' to control access. That means tickets and attachments can be made visible only to specific users, departments, or project teams. These levels are visible in the ticket's source code. Since our Tesla Jira tickets are saved as archived web pages, we can check their source code. In nearly all of them, we find this fragment: Security Level – Viewable by All value = Viewable by All Users.

It sounds like a breakthrough. We take it as confirmation: Jira tickets at Tesla were visible to anyone inside the system. But a closer look at the source code undermines our assumption. The setting doesn't govern the entire ticket – only the comment section. When Tesla employees write a comment on a ticket, they can choose who's allowed to read it. The default is 'Viewable by All Users.'

It's one of those moments in the Tesla investigation where everything seems to grind to a halt. How can we prove Lukasz's account if we still can't talk to Tesla insiders – the only people who could truly confirm the authenticity of the data? The alternative route was through technical analysis. But the evidence we hoped for has just slipped away.

Viewable by All

We go through all 1,700 Jira tickets one more time. We examine how they're displayed in the browser, dig into the source code, and search for any further signs of access control. And finally – after searching through the digital haystack – we find them: a small number of tickets that actually include a 'security level' setting. Not just the comments – but the entire ticket. Even better: these files list access permissions as 'all,' 'everyone,' 'unrestricted,' or 'available to all users.'

A small detail – but a crucial one for our investigation. From this, we can draw three key conclusions. First, Tesla does in fact use 'security levels' to manage Jira access. Second, in the overwhelming majority of tickets in the Tesla Files, no such level is defined. And third, where it is used, the assigned user groups are strikingly broad – even in tickets covering topics like internal planning in car-body manufacturing or sensitive server configurations.

Jira offers another form of access management: user profiles can be assigned to groups or project roles that determine their permissions. Our dataset contains dozens of these user profiles from Tesla's Jira system. They show, for example, that Joe Ward, Tesla's vice president for Europe, belongs to 73 groups. A service technician appears in 107 groups. Lukasz – a low-level technician – was a member of 137. These weren't just warehouse or service teams, but also labels like 'All-Global-Admin' and 'All-EMEA-Admin.' Admin stands for administrator, a user with elevated privileges. EMEA refers to Europe, the Middle East, and Africa.

We begin to wonder: is Tesla being overly generous – or simply careless? Are such sweeping permissions standard practice in a global tech company?

We also find evidence that Lukasz's access went beyond read-only. He was able to edit and close tickets. This, too, points to a technically differentiated rights structure. But was it Tesla's intention to give someone at Lukasz's level access to sensitive material across departments and continents?

Soon we find another indication of risky access management: an email from an employee in Las Vegas, dated September 20, 2018. The subject line reads: 'Permissions Update *Action Required*' She sounded the alarm after an internal assessment had uncovered 'large amounts of confidential & sensitive data accessible and visible by a very large number of Tesla employees.' She warned that groups like 'Confluence users,' 'Employees,' and 'Everybody' had been granted access – even though each included more than 25,000 users. She urged her colleagues to take immediate steps: 'Please take the time to review some next steps to improve our security posture across the company.'

It sounds like the problem had been clearly identified – six weeks before Lukasz began working at the Tesla service center in Norway.

It's hard to believe that her warning simply faded away, without triggering a security audit or anyone bothering to fix the gaps. Then we find an email from Elon Musk himself, sent to employees on June 17, 2018. Subject: 'Some Concerning News.' Musk wrote that an employee had conducted 'quite extensive and damaging sabotage to our operations.' This included 'making direct code changes to the Tesla Manufacturing Operating System under false usernames and exporting large amounts of highly sensitive Tesla data to unknown third parties.' The motive? Disappointment over a missed promotion – or maybe something more. Musk's message: 'Please be extremely vigilant.'

Again, we're stunned. On the one hand, there's a dramatic warning from Musk about the importance of data security. On the other, there are clear signs that in practice, the company took a remarkably lax approach to safeguarding internal information. The two simply don't add up.

When we began this investigation, our first goal was to verify the authenticity and origin of the data. By now, we've checked both boxes. We're convinced the data is real. It may be chaotic and inconsistent, but it's also unmistakably coherent.

In thousands of spot checks, we haven't found a single false entry. While some on the team were buried in Jira, others examined thousands of separate files. We discovered cross-references, documents that built on each other, and overlapping staff records. Could AI have generated all of this? Not impossible – but highly unlikely.

We now believe we understand the root of Tesla's data protection problem. And the files lay it all bare.

Fault Lines

When we lay out our findings to the editor-in-chief and legal counsel, their response is cautious – measured, but far from convinced. We sit shoulder to shoulder in a glass-walled conference room on the sixth floor of *Handelsblatt* headquarters, laptops open in front of us. Colleagues passing by on their way to the break room can see our tense faces – but can't hear a word of what we're discussing. Apart from Lukasz, no one outside this room knows about the project – and that's exactly how we intend to keep it. It's too sensitive. We have a source to protect.

Of course, word spreads after our secret meeting. A colleague later pokes his head into our office and asks, 'So, what have you investigative types stirred up this time?' We laugh. 'Oh, you know . . . just the usual.'

It's a massive understatement.

Can we truly rule out the possibility that a handful of fake files were deliberately planted among thousands of authentic ones? The Tesla story, our legal counsel warns, has the potential to put this entire publishing house at risk. He's referring to so-called SLAPP lawsuits – Strategic Lawsuits Against Public Participation. They are designed to intimidate critics and silence them through the threat of crippling legal costs. Musk is notorious for this tactic. Media outlets are frequent targets. The threat of multimillion- or even billion-dollar claims is meant to suppress coverage.

There are now laws recognizing such lawsuits as abusive – but the risk still isn't zero. Sebastian Matthes, editor-in-chief, supports our investigation, but has one more request: bring in external IT experts to assess whether the files may have been manipulated.

We choose the Fraunhofer Institute for Secure Information Technology (SIT), one of the world's leading centers for digital forensics. On February 22, 2023, we drive to their campus in Darmstadt with our legal counsel, Peter Koppe. There, we meet with Martin Steinbach from the Media Security and IT Forensics division as well as two of his colleagues. Their specialty is authenticity analysis: searching for traces of manipulation inside digital files.

Steinbach and his team are the first people outside our newsroom to see the material. It makes us a little nervous. To be safe, we ask the outsiders to sign a nondisclosure agreement. They don't hesitate. Once confidentiality is agreed, we walk them through the background, the nature of the dataset, and our findings so far. They agree to take on the task.

There is a caveat, the experts warn us: if the data has been professionally falsified, they may not be able to prove it. IT forensics can only guarantee the integrity of data when it has been collected using certified forensic methods. That isn't the case here. Lukasz didn't extract the data in any structured or secure way. He downloaded it freely. His screenshots and videos, the experts explain, are of limited forensic value. They could have been altered using simple tools. Some contain no metadata at all – making verification difficult or impossible.

On the drive back to *Handelsblatt* headquarters in Düsseldorf, we run through every scenario again and again. What if Fraunhofer can't help us? What if they find anomalies? That would kill the investigation. Weeks of work – abandoned.

The thought is hard to shake off.

Soon afterward, we get the green light. Fraunhofer reports that its analysts found no indication that 'the dataset did not originate from Tesla IT systems or Tesla-related environments.' Among other things, they analyzed GPS coordinates in the image metadata. The locations all make sense. Many point to areas near Tesla facilities – in France, Japan, and the United States. The experts confirmed that Tesla's Jira system likely uses a differentiated access rights model. But they also found that Lukasz – and many others – had been granted unusually broad access.

The direction is now clear. We're moving forward on the assumption that the data is authentic, and that the whistleblower is telling the truth about its origin and the company's lack of data security. Still, everything we publish must be verified – once, twice, three times. The Fraunhofer report fires the starting gun for the next phase of our investigation.

The Cult and the Cracks

Our optimism fades fast. The silence is deafening. Out of more than 200 people we reach out to, hardly anyone responds. When a message does come back, or someone picks up the phone, it's usually just to say no. Employees point to the airtight confidentiality clauses they signed, forbidding them from speaking about anything that happens inside the company. Former staff worry about lawsuits if they talk about their time at Tesla. We offer source protection. Anonymity. It makes no difference. Even customers who've had bad experiences stay cautious. Too loyal. Too afraid. Tesla remains a black box.

So what now? We decide to leverage what we have. We don't just know who we're calling – we know where they work, what they earn, in some cases even their medical history. And so, we turn to something deeply human: curiosity.

When we call up a former Tesla manager, he's polite but distant. Says he doesn't talk about former employers – especially not this one. We gently push. We read him his employee ID. Then his job title. Then his home address. He doesn't object. Then we tell him the amount of his final paycheck. A pause. Papers rustle. 'To the cent,' he says. Then he starts talking.

This approach works with current employees, too. A sales rep in Germany shuts us down at first. But once we offer a glimpse of what we know, the wall crumbles. For 90 minutes, he tells us what it's like inside his Tesla dealership – the pressure, the dissatisfaction, the way his manager keeps the team on edge. We listen, transfixed.

There's a daily call, he says, where every store manager has to report how many cars they've sold that day. Sometimes, he feels like a character in the movie *The Wolf of Wall Street*. If the store hasn't sold a single car by late afternoon, the manager storms through the showroom, yelling that he isn't going into that call with a zero on the board. Nobody is allowed to leave until that number starts with a one. Some nights, staff have to stay at the dealership past 8 p.m.

As we speak, the Tesla employee sends us screenshots – messages with his boss.

They show how internal complaints are shut down. When someone challenges the gap between recruitment promises and the harsh reality of the job, the manager responds by pointing to the 'mindset of flexibility and adaptability' expected of all employees. This is Tesla. Not just a job. A mission.

'It's like a cult,' the man says on the phone. 'A totally crazy world.' It feels like a pressure-cooker sales floor. Those who perform are kings. Those who don't? They're nothing. Before hanging up, the Tesla employee promises to connect us with other frustrated colleagues. Bingo.

What works with employees works with customers, too. Naturally, they're suspicious when a stranger calls them, asking about their car. But they start listening when that stranger names the exact issue they reported, the date they visited a service center, and the precise distance they had driven their Tesla by the time of the reported incident.

One British customer, who had brake problems and was later involved in an accident, writes: 'Sorry to be suspicious but I was wondering if you could tell me your source for my VIN number and odometer?' He might be interested in telling his story, he explains, but first he would like to understand the motivation for our work – and, as he puts it, 'whether what I experienced is not as rare as I was maybe led to believe.'

We tell him we have access to internal Tesla data and

that his name appeared in a table with hundreds of similar cases. That we're verifying what we've found, and seeking as many firsthand accounts as possible. He marvels at this. We can tell: this man is not a happy customer.

Five days later, he sends us a detailed account of the crash – and how Tesla handled it. He even attaches dash-cam footage taken just before the collision. The video shows his car on a busy London road, following at a safe distance – then plowing into a braking truck. The customer ends his message with: 'I am happy to try and help you further and also look forward to hearing what you have found from other Tesla owners.'

And so, piece by piece, we begin to assemble the Tesla puzzle. We track everyone we contact in a master spread-sheet. Red means rejection. Green means conversation. Every day, more names turn green. Some of the red ones turn green, too. Every green square is a tiny crack in Tesla's black box. Every voice brings light into the dark, illuminating another corner of Elon Musk's empire.

We collect the interview notes in a shared folder. Each conversation yields new insight; many become bridges to others. Employees tell us about what hiring looks like – and what getting fired feels like. Frustrated customers show us their email exchanges with the company. They describe service centers where they weren't taken seriously. Those who went to court send rulings and expert opinions.

The volume and depth of information is amazing. All this from inside Tesla's black box! Just weeks ago, it would have been unthinkable.

It's time to start thinking about publication. Of course, we'll need to write about the massive security leak Lukasz discovered. But to show just how severe that breach is, we want to follow the first article with a second one. There's no shortage of topics. In the end, the choice is obvious. Our opening story will focus on Tesla's most ambitious – and most dangerous – project: Autopilot.

Confirmation by Threat

Building sleek and speedy electric cars is no longer enough, says Elon Musk. Competitors have long caught up, with their own electric models. Autonomous driving is the core promise around which Musk has built his company. And yet, Tesla has never delivered a truly self-driving vehicle. Still, Musk keeps repeating the same claim: that his cars will soon drive entirely without human help. The central question: Is Autopilot really as advanced as he says?

The Tesla Files suggest otherwise. We've found spreadsheets filled with complaints from customers whose cars had accelerated or braked without cause. We want to know more. The problem: the files list only vehicle IDs, not names. So we begin merging datasets, layering one over another until overlapping details emerge – often enough to identify a skeptical customer willing to talk. We go case by case.

Meanwhile, we press ahead with our data-privacy investigation. The Tesla Files include letters Lukasz sent to various authorities. He first contacted the U.S. highway regulator, the NHTSA, then agencies in Norway, where he worked for Tesla. When none replied, he escalated to the U.S. Securities and Exchange Commission (SEC) – the federal agency responsible for enforcing securities laws and protecting investors – and eventually to the German Federal Data Protection Commissioner.

We realize that, long before we ever heard from him, Lukasz did us a big favor. If German authorities were informed about safety issues at Tesla in the past, now we can ask what action – if any – those authorities have taken. That alone would be newsworthy. Violations of privacy laws can be expensive: fines can reach up to 4 percent of a company's global annual revenue. For Tesla in 2022, that would mean $3.26 billion. But when we send our inquiries, disappointment sets in fast. The replies are as brief as they are clear: nothing. The agencies have done absolutely nothing.

We want to know why. Lukasz shares a reply he received from the Federal Commissioner: 'My jurisdiction is limited to oversight of federal public agencies, except for companies providing telecom or postal services, or those covered by national security law,' the official wrote in German. Compliance for all other companies 'falls under the state-level data protection authorities for the non-public sector. Jurisdiction depends on the location of the responsible office or its only German branch.'

Lukasz didn't know what to make of that – understandably. When we explain it to him, he reacts immediately. He forwards his files to the state authority in Brandenburg, where Tesla's only European factory is located.

When we follow up, we get a prompt response: 'The commissioner has received credible indications of possible data privacy violations by Tesla,' a spokesperson for state data officer Dagmar Hartge tells us. The matter concerns 'sensitive employee data,' which may have been 'broadly accessible due to insufficient access restrictions within the company.' If true, the case would be 'particularly serious, given the large number of people affected worldwide.'

Hartge says her office will investigate what exactly the Tesla Files contain and that, due to the international dimension, she has informed Dutch regulators as well; Tesla's European HQ is in the Netherlands.

What we don't know yet: just weeks after our first story, over two dozen regulators across Europe will begin investigating the Tesla Files.

Now it's time to contact Tesla. The same journalistic principle applies: if there are allegations, the company must be afforded an opportunity to respond. By now, our article has grown to over 5,000 words. We go through each line: is there a question Tesla needs to answer? A claim they should have the chance to address?

Our legal team does the same. General counsel Peter Koppe reviews every word. He brings in Professor Roger Mann, one of Germany's top media lawyers. Mann, in turn, consults an expert in the U.S. Again and again, we remind each other: This is Elon Musk we are dealing with here. The richest man in the world. A man known to react with lawsuits – anywhere, against anyone. What if he sues *Handelsblatt* in New York, Texas, and California? Or even in London, Tokyo, or Beijing? He has the resources. More than anyone.

And so, the Tesla story becomes a top-level operation. Editor-in-chief Sebastian Matthes reads multiple versions. The board is informed. Even our publisher, Dieter von Holtzbrinck, is in the loop. After all, it's his money that might be on the line. But Holtzbrinck is old-school. The decision, he says, rests with the newsroom. If everything is solid and every legal angle covered, we go to print.

So, on May 10, 2023, we confront Tesla. In newsroom jargon, that means sending formal questions on the subjects we're reporting on. Most companies have PR departments for this. Tesla doesn't.

Journalists have long complained that Tesla doesn't respond to media inquiries. The company disbanded its PR team in 2020. Musk later confirmed on Twitter: 'Other companies spend money on advertising & manipulating public opinion, Tesla focuses on the product. I trust the people.'

After acquiring Twitter in 2022, Musk did the same there – gutting the PR team. For a while, sending a press question to Tesla meant getting one response: a poop emoji.

We don't expect a reply. But we have to try. There's just one problem. Who takes questions if there's no media department?

Again, the Tesla Files help. We search employee lists for legal staff, comms managers, Musk's assistants, and his direct reports. For good measure, we *cc* Musk himself – thanks to the Tesla Files, we have three of his email addresses.

And so, Musk and his top team receive 65 questions from Düsseldorf. About data privacy. About Autopilot. We ask if Musk's social security number in the Tesla Files is real. How Tesla protects user data. Whether they'll cooperate with regulators. Why so many cars reportedly brake or accelerate on their own. Why Musk keeps promising full autonomy without delivering. We give Tesla seven days to respond.

May 17. The *Handelsblatt* newsroom is buzzing: our global correspondents' summit is in full swing. Reporters from around the world are here. And then, in the middle of the night, it arrives: An email from Tesla. No emoji. A real response. Attached: a letter from Tesla's legal department. It's a gift.

Tesla attorney Joseph Alm, writing from Austin, Texas, lectures us. *Handelsblatt*, he says, received the data from 'a disgruntled ex-employee' who 'misused his access as a service technician to exfiltrate information in violation of his signed non-disclosure agreement, Tesla's data management policies and practices, and EU and German law.'

It's a mouthful, but the message is clear. Tesla threatens legal action. 'We will also cooperate with criminal authorities in this matter,' the letter says. 'The possession of such data itself without a proper justification breaches, among other things, data protection law. [. . .] Any such sensitive data in your possession also requires you to protect it carefully against further misappropriation. [. . .] As you know, use of illegally obtained data for media reporting is not allowed absent exceptional circumstances.'

Mishandling the material, Alm warns us, could expose *Handelsblatt* to liability for 'trade secrets, data protection law, and handling stolen data, among other things.' Anyone with such data, he insists, must 'protect it carefully against further misappropriation.' Tesla demands we delete the files – and confirm the deletion.

Sönke is the first to see the email. He forwards it before dawn. At the office, we greet each other with high-fives and hugs. We take Tesla's response as a badge of honor. They haven't answered our questions – but they haven't denied the authenticity of the files either. Nor have they claimed that anything we have is false.

When we meet Peter Koppe, our general counsel, he gives us a perfect smile. This isn't the answer he expected, but it's the best he could have hoped for. No further questions, your honor. At the *Handelsblatt* summit, the day flies by. The investigative team celebrates into the night.

We did it.

CHAPTER 4

The Autopilot Illusion

In the end, the moment we've spent half a year working toward feels strangely unspectacular. It's the late afternoon of May 25, 2023, and we're gathered in the newsroom on the sixth floor of the *Handelsblatt* publishing house. We have spent tense days filled with last, truly last, and absolutely final change-requests from the legal department, and debates about the right cover for our story. We pace around the building, driven by the fear that we've overlooked something. That we've missed a legally relevant detail. That something might still stop us.

By 6 p.m., we're at the central desk with cold drinks in hand. Every caption, no matter how short, has been checked three times. In front of us sit the CVDs – the chiefs on duty, forming the newsroom's command bridge. They are the last to review our article before placing it on the website. They also send the push notifications to all users of the *Handelsblatt* app. When our phones vibrate, our anticipation roars.

It feels surreal to see the work of so many weeks now displayed on our phones. One last check – everything looks as it should. Then we toast. On the nearby table, a laptop connects us with colleagues in Berlin, the U.S., and Japan. We thank the designers, layout team, and everyone else who helped bring the project to life – and now stay to celebrate with us. For a while, we watch the climbing bars in the analytics tools we use to track online readership. Interest is surging.

We share a few last impressions from the investigation, speculate on how – and whether – Tesla might react, and discuss next steps. Then we head into the evening. As we leave the building and climb into our cars, onto bikes, or onto trains, the first reactions start coming in. Colleagues congratulating us on the research. Sources thanking us for the articles. And Musk fans, accusing *Handelsblatt* of waging a smear campaign on behalf of the German auto industry. One will go on to comment under every one of our articles in the months ahead: '*Tesla wins – Handelsblatt loses.*'

From Tesla, we hear nothing. Journalists quickly learn: comment sections are best consumed in small doses. Still, that evening, many of us are glued to social media. Standing in playgrounds, lying on sofas, refreshing our feeds on Twitter and LinkedIn every few minutes. We wonder: will Elon Musk respond to the *Handelsblatt* article? But the

Tesla CEO stays silent. We take that as final confirmation of our reporting.

The next morning, we reach for our phones as soon as we wake up. A Google search for the latest Tesla news brings more than just our report. Overnight, our investigation has gone global. Newsrooms across Europe are covering it. There's keen interest from Australia, China, and India. In the U.S., the story spreads quickly – the *Los Angeles Times*, *Wired*, CNN, CNBC, *The Verge*, and countless other outlets quote what the Tesla Files reveal about Elon Musk's most important promise. Each headline reads like another crack in the polished image Musk has built over the years. Many even use our headline – a customer complaint found in the Tesla Files: 'My Autopilot almost killed me!'

The quote stems from one of the Excel spreadsheets documenting customer feedback. These files form the core of our Autopilot story. Lukasz exported them from Tesla's 'Toolbox system,' the main communication platform between customer service and other departments. It's where Tesla logs incidents reported by customers, along with diagnostics and repair procedures. Each complaint list carries a file name with a number tied to a specific issue. 26977 stands for 'Customer reports collision / accident / crash.' 29034: 'Request from Authorities (e.g. Police, Law Enforcement, Legal representative, Public Prosecutor) for vehicle data.' There are codes for heater malfunctions, others for broken cameras. One spreadsheet lists only missing games in the Tesla Arcade. Another logs hundreds of drivers who accidentally ordered Autopilot features and want to reverse them.

Two codes matter most to us. 27973: 'Customer claims an unintended acceleration event occurred.' 55538: 'Automatic Emergency Braking (AEB) operated for unknown reason.' In the industry, they're known as 'unintended acceleration' and 'phantom braking.'

These aren't isolated incidents. In the Tesla Files, we find an internal presentation from May 2018. A Tesla engineer lists ten particularly troublesome error patterns. Among the most serious: unintentional acceleration and braking – exactly what our article focuses on. These impair 'the safe operation of the vehicle,' the slide notes. Beside it, an alert: 'Hazardous Without Warning. Direct Risk to Customer Safety.'

The Tesla Files contain more than 2,400 complaints about unintended acceleration and more than 1,500 braking issues – 139 involving emergency braking without cause, 383 phantom braking events triggered by false collision warnings. More than 1,000 crashes are documented. A separate spreadsheet on driver-assistance incidents where customers raised safety concerns lists over 3,000 entries. The oldest date from 2015, the most recent from March 2022. In that time, Tesla delivered roughly 2.6 million vehicles with Autopilot software. Most incidents occurred in the U.S., but we also find complaints from Europe and Asia.

Customers describe their cars suddenly accelerating or braking hard. Some escaped with a scare. Others ended up in ditches, crashed into walls, or collided with oncoming vehicles. 'After dropping my son off in his school parking lot as I go to make right hand exit it lurches forward suddenly,' one complaint reads. Another says: 'My Autopilot failed/malfunctioned this morning (car didn't brake) and I almost rear ended somebody at 65 mph.' A third reports to Tesla: 'Today, while my wife was driving with our baby in the car, it suddenly accelerated out of nowhere.'

Braking for no reason causes just as much distress. 'Our car just stopped on the highway. That was terrifying,' a Tesla driver writes. Another complains: 'Frequent Phantom braking on two-lane highways. Makes the Autopilot almost unusable.' Some report their car 'jumped lanes unexpectedly' causing them to hit a concrete barrier or veered into oncoming traffic.

Some of the quotes are so stark that we opt for a narrative device. We split our article into eight sections. At the start of each, we place three quotes from the spreadsheets. These fragments are meant to strike the reader like a refrain. Beneath them, the question that runs through everything: *Just how dangerous is Tesla's Autopilot?*

To find out, we spent the past weeks contacting dozens of Tesla drivers across several countries. By combining different spreadsheets, we were able to match specific incidents to names using vehicle ID numbers. All of them confirmed the information about them in the Tesla Files: that the incident happened, that the mileage Tesla recorded was correct, and that they had filed complaints. In personal conversations, they shared their experience with the Autopilot. Some gave us access to their communication with Tesla. Others sent us videos of the incidents.

Unfit for the Road

One of these customers is Thomas Karl. The Swiss was an early fan. He says he bought the 38th Tesla in Switzerland back in 2013. Karl was enthusiastic about the vehicle and recommended it to friends and acquaintances. But his next Tesla caused problems. The car could accelerate from 0 to 100 km/h in 3.8 seconds – but just as abruptly, it would decelerate. Karl filed multiple complaints with Tesla because the car would suddenly brake hard without warning. 'I keep experiencing phantom braking – it's not only annoying and has nearly caused a divorce (if my wife is in the car, I'm only allowed to drive manually – which is great for a vehicle that costs over CHF 100,000!), but it's also extremely dangerous,' Karl wrote to Tesla on January 13, 2021. On the highway, no one expects 'a car ahead to suddenly brake just before a tunnel entrance, simply because there's a traffic light mounted above the entrance that's turned off but is recognized by the Tesla as red!'

A Tesla employee asked for the time and location of the incidents 'to better interpret the vehicle's response.' Before Tesla could offer any explanation, Karl complained again. In an email dated January 29, 2021, he wrote: 'Another phantom braking incident today – it's still a horror show and extremely dangerous. Today, emergency braking was triggered even though the oncoming vehicle was clearly in its own lane and there was another car behind me.'

On February 5, 2021, Tesla got back to Karl. The company cited 'local conditions' as the cause of the braking events. The system braked because 'the minimum lane width was barely met' on the roads where the incidents occurred. The car interpreted the path as a single lane and saw no room to evade, so it braked in response to oncoming traffic. Tesla assured him that the systems were 'behaving absolutely normally, that there is no fault with the vehicle, and therefore no defect.' Nevertheless, Karl's vehicle would transmit data from 'such driving situations' to Tesla so they could be 'virtually practiced and improved through driving simulations and machine learning.' It could not be ruled out 'that your vehicle may handle such situations better with future software updates – or even flawlessly.'

In a follow-up email on February 15, 2021, the Tesla employee added that when driving his own Model S on narrower or complex roads, he preferred to take over manually, because 'the assistance systems are too jittery in such situations.' On the highway, the system worked 'relatively well,' he said, although the car would abruptly brake when another vehicle entered the access road 'too quickly or far too slowly,' and he would 'usually have to intervene.' He added, 'With that knowledge, you can drive quite proactively and encounter few surprises. Though I have to say, with every update, the vehicle handles one or two situations better.'

For Karl, that was little consolation. He wanted to feel safe in his luxury vehicle – immediately, not at some point in the distant future. But his Tesla didn't make fewer mistakes; it made more. 'That's enough – yesterday my Tesla braked three times again, one of them so violently on the highway that it was pure luck there was no car behind me,' Karl wrote to Tesla on April 7, 2021. 'I'm amazed you manage to sell so many vehicles despite these problems and your arrogance.' Out of frustration, he put his Tesla up for sale. 'This is a car that's not fit for use.'

A Tesla employee offered to join Karl in the car for a test drive. Karl agreed. Together, they drove along routes where problems had occurred. In an email dated May 3, 2021, the Tesla employee concluded that the Autopilot's faults were 'purely due to software.' In some cases, the sudden braking was even intentional – 'for safety reasons.' In one instance, Karl's car likely braked due to the narrow lane: 'Unfortunately, we can't explain why it does that here.' His advice to Karl: just keep your foot on the accelerator. 'That's what I do these days too – and I manage.'

Karl did not manage. 'Good day, gentlemen – do you believe me when I say I'm reaching the end of my rope?' he wrote on July 26, 2021. His Tesla had just performed a full emergency stop on the Swiss Autobahn A3 between Flums and Sargans after overtaking a vehicle, 'so abruptly it was downright frightening.' He warned Tesla: 'This car is simply too dangerous for the road.' Again, no solution was found. Karl's final complaint is dated September 29, 2021: 'Unfortunately, I must again report unreasonable braking in my Model S,' he wrote. 'On the stretch between Rapperswil–Jona and Eschenbach, I had another three braking events, all of them completely unjustified.' He protested that the situation remained 'unacceptable' and that his wife was becoming 'increasingly allergic to the vehicle.' He called on Tesla to buy the car back. The company refused. When we spoke to Karl in early 2023, he had lost all faith in Tesla and sold the vehicle privately. He now drives an electric model from Skoda.

Another example of Tesla's technical issues is the experience of Manfred Schon. The software developer and former employee of the German automotive supplier Bosch contacted us about an incident from October 2019. At the time, Schon was driving his Tesla along the M14 road between his hometown of Northville and Ann Arbor, Michigan, on a flat, straight road. Another vehicle was about 100 meters ahead when he noticed brake lights in his lane – but no signs of congestion. Schon says he eased off the accelerator but did not brake. 'All of a sudden, the car slammed on the brakes – as hard as you can possibly imagine. It wasn't gradual at all.' Seconds later, his Tesla stopped in the middle of the lane, Schon's seatbelt pulled tight across his chest. Then the car behind slammed into him.

Looking back, Schon was angry at himself. It wasn't the first time he'd had issues with his Tesla. But he had hoped that the Autopilot would improve with future software updates. After the crash, he wrote to Tesla on October 28, 2019. The unintended braking was 'no longer acceptable' and represented 'a safety hazard,' he complained. He never received a response – not even an acknowledgment of receipt.

Still, Schon kept driving the high-risk vehicle. During service visits, he mentioned the incident, but technicians brushed it off. Probably a software issue, they said. He should just wait – updates would fix it over time. After many more phantom-braking episodes, on March 18, 2021 Schon managed to obtain the email address of a Tesla employee. Under the subject line 'Model S Safety Issue,' he described the unwanted braking in detail. Tesla asked him to bring the vehicle to a service center, where it was kept for several weeks. When he returned to pick it up, he was told no fault had been found. The company said there was nothing more it could do.

Schon gave up. He suggested Tesla buy back the car at market value – minus the cost of the Autopilot package.

No answer. Selling it himself was out of the question. He couldn't do it – not in good conscience, he told us. What shocked him most was Tesla's attitude: 'This complete lack of concern given the seriousness of the safety problems.'

A physician from Southern California, who prefers to remain anonymous, also tells us that her Tesla responded dangerously in several situations. She was thrilled when she got into her first Tesla in 2017. But after just two months, the problems began. Sometimes the car would accelerate instead of braking, then brake when it should have accelerated. 'It was so frightening,' she says. She reported the incidents to Tesla immediately. Once, she was even injured. In fall 2021, while trying to turn around in a parking lot, her Tesla suddenly accelerated like a race car.

'I tried to steer, but crashed into a concrete post,' she recalls. 'It toppled over, but the car didn't stop. I hit the next post. The airbag deployed, and I was in shock.' Tesla immediately realized something had happened and called her. She was covered in bruises. Later, Tesla blamed her for the crash, but she insists she never touched the accelerator. She requested a replacement vehicle – being seriously afraid of her car now. Tesla didn't replace it, only repaired it. Today, she pays a higher insurance premium because of the incident. She tells us: 'If there's ever a class-action lawsuit, I'm in.'

Deadly by Design?

Not all autopilots are created equal. In the automotive industry, five levels of autonomous driving have become standard. Level one includes driver-assistance systems like lane-keeping and emergency braking – now common for many brands. Level two refers to systems that can temporarily drive on their own, taking over steering, braking, and acceleration, but full control remains with the driver, who must keep their eyes on the road and be ready to intervene at any time. With level three, that changes. In certain situations – on highways, for instance – the system takes full responsibility. The driver may look away from the road, though they must be ready to take over when prompted. At level four, vehicles operate without human input in specific environments and under certain weather conditions. Level five is full autonomy: a car that drives itself in any environment.

Tesla's Autopilot most closely matches level two – a system designed to assist, not replace, the driver. Customers in both the U.S. and Europe are legally required to keep their hands on the wheel and eyes on the road at all times. That's what it says in the manuals of Musk's electric cars. Still, Tesla marketed the most expensive version of its autopilot for years as 'Full Self-Driving', priced at $15,000. Critics accuse Musk of deceiving customers.

One of them is Tesla co-founder Martin Eberhard. The engineer believes it is 'dangerous' to release an autonomous vehicle 'onto the road before it is one hundred percent safe and reliable.' Apple legend Steve Wozniak – whose data appears in the Tesla Files – said in an interview that Musk personally convinced him to buy a Tesla, promising it would be able to drive itself across the country by the end of 2016. 'I actually believed those things, and it's not even close to reality,' Wozniak told CNN in 2023. 'And boy, if you want a study of AI gone wrong and taking a lot of claims and trying to kill you every chance it can, get a Tesla.'

Vivek Wadhwa tells a similar story. The California-based entrepreneur and author met Musk in 2013. The Tesla CEO personally persuaded him to buy a Model S. Excited, Wadhwa upgraded to a Model X in 2016 to access the latest Autopilot. That's when the problems began. His Tesla once drove itself into his garage, he says. But that was just the beginning. As a fan of the brand, Wadhwa allowed PBS to interview him in his car in 2017. He had planned to demonstrate how well Autopilot worked. But with the camera rolling, his Tesla came within inches of rear-ending the car ahead – Wadhwa hit the brakes at the last second. Today, he is a vocal critic. 'Elon keeps pushing a lie. People are dying because of Tesla's faulty technology,' Wadhwa says. 'And Elon is trying to get away with it.'

Even more dramatic is Dan O'Dowd, a billionaire from California. His company develops software for the mobility sector – making him a direct competitor of Tesla. O'Dowd founded The Dawn Project, an initiative aimed at uncovering safety flaws in other manufacturers' systems. Musk, Tesla, and Autopilot are its primary targets. The campaign frequently grabs headlines with bold stunts. In February 2023, O'Dowd spent $7 million on a 30-second ad during the Super Bowl; more than 113 million people in the U.S. watched the NFL championship. The clip showed allegedly self-driving Teslas breaking traffic laws, hitting a stroller, and running over a child-sized dummy on a pedestrian crossing. A voice-over warned: 'Tesla's Full Self-Driving endangers the public.'

To find out whether these claims are exaggerated, we head to the rural district of Landsberg am Lech in German Bavaria. There, Jürgen Zimmermann has converted an old cowshed into a workshop. Up to 700 Teslas a year roll onto his car lift, he says. The automotive technician is something of a celebrity in the Tesla world. He films himself inspecting the cars, removing wheels, cursing at drive shafts. His YouTube videos are watched by hundreds of thousands. Zimmermann calls it 'unreal' to drive a Tesla. But for all his admiration, he's not blind to the flaws.

Zimmermann has a theory about the sudden-acceleration issue. Until recently, Tesla relied not only on camera input, but also on a low-resolution, front-facing radar. The system asked one binary question: 'Is there a car in front of me – yes or no?' Nothing in between, Zimmermann says. On a tight bend, the radar might lose sight of the vehicle ahead – and the car would accelerate.

In May 2021, Tesla removed the radar sensor. Zimmermann believes the problem is likely to diminish in newer models. Data from the Tesla Files seems to support that. Since switching to camera-only Autopilot, instances of unintended acceleration appear to have declined.

But phantom braking incidents have increased. Zimmermann suspects faulty camera calibration may be the cause. He's about to test the idea himself – by loading his trunk with weights. When we say goodbye, he promises to keep us posted. A few days later, he does: his theory didn't hold. He now suspects phantom braking is primarily a software issue. Zimmermann's new hypothesis: Autopilot mistakes shadows or other harmless objects for obstacles.

In principle, the camera system works like this: a central computer processes the flood of images using artificial intelligence. Over the past decade, AI has made significant advances in image recognition. But perfect accuracy remains out of reach. In complex environments – especially cities – detection rates plummet. Every parked car presents a question to the AI: How long has it been there? Is it about to move? Cameras also struggle in tunnels, at night, and in backlight. Snow and rain pose additional challenges.

That's why nearly all other carmakers use additional sensors such as radar or lidar. These scan the environment using electromagnetic waves or laser beams and can provide information when cameras fail. But Musk categorically rejected putting sensors back into the vehicles. His reasoning: 'In my view, it's a crutch.' Instead, the answer is supposed to be in data. According to a presentation at Tesla's Investor Day in March 2023, the system has been trained with 14,000 videos 'from our own fleet' showing 'braking due to some parked car.' The goal of eliminating phantom braking is still out of reach.

Tesla, of course, is not the only automaker struggling with autonomous driving. According to a McKinsey analysis, automakers and venture capitalists invested $106 billion between 2010 and 2020. The results have been underwhelming across the industry. But Tesla remains, by far, the carmaker that promotes its Autopilot most aggressively. And Musk has long claimed that his system in particular is practically flawless.

Musk's Endless Promise

More than a decade has passed since the Tesla CEO declared a revolution for the roads. 'I think we'll be able to achieve true self-driving in five or six years,' Musk said in October 2014. What he envisioned was a journey 'where you could literally get in the car, go to sleep and wake up at your destination.'

In January 2016, Musk declared the race for the future of human transport was over: 'I really consider autonomous driving a solved problem.' In April 2019, at Tesla's Autonomy Investor Day in Palo Alto, he suggested 'the fundamental message consumers should be taking today is that it's financially insane to buy anything other than a Tesla. It'll be like owning a horse in three years.' The Tesla CEO added: 'By 2020 we will absolutely have a million robotaxis on the road.' Another prediction that day: 'Probably two years from now we'll make a car with no steering wheels or pedals.'

None of those predictions ever came true, even though in June 2022 Musk called the Autopilot a matter of survival. 'The overwhelming focus is on solving full self-driving. That's essential,' he told the fanclub Tesla Owners Silicon Valley in an interview. 'It's really the difference between Tesla being worth a lot of money or worth basically zero.'

In April 2023, Musk made a new promise. The trend was 'very clearly toward full self-driving, toward full autonomy,' Musk said during Tesla's Q1 earnings call. Tesla, he said, was making rapid progress. His forecast: 'I hesitate to say this, but I think we'll do it this year.'

Digging through Musk's past statements is straightforward archival work – but turns into one of the most striking moments of our investigation. We hadn't realized how brazenly Musk had repeated the same promise, over and over, for years – with nothing to show for it.

We ask ourselves whether a CEO of any of the 40 companies listed in the German stock index DAX could get away with something like this. What if the heads of Volkswagen or Mercedes repeatedly unveiled products that never reached market? How long would it take before they were accused of stock manipulation? Wouldn't authorities have stepped in long ago – if it wasn't Musk?

There are people who deliberately spread false or misleading claims to inflate a stock's value and sell before it crashes. In industry slang, that tactic is called a 'pump and dump.' Isn't Tesla's CEO walking that very line with more than a decade of recycled Autopilot hype?

Anyone looking into Musk's broken promises will eventually stumble across a 2016 promotional video. It shows a Model X apparently driving itself autonomously through California. At the start, Tesla displays white text on a black screen: 'The person in the driver's seat is only there for legal reasons. He is not doing anything. The car is driving itself.'

The soundtrack: 'Paint It, Black' by the Rolling Stones. On October 20, 2016, Musk tweeted: 'Tesla drives itself (no human input at all) thru urban streets to highway to streets, then finds a parking spot.'

His followers were ecstatic. 'Wow, absolutely fantastic,' one comment read. Others flooded his mentions with praise, calling him as 'a genius' and 'one of the greatest minds and revolutionaries of our time.' The tweet earned over 17,000 likes and more than 11,000 retweets.

Seven years later, the world learned that the claim of full autonomy wasn't exactly what it seemed.

In January 2023, Ashok Elluswamy – Tesla's head of Autopilot software – was called to testify in court over a fatal accident. On the stand, he made a startling admission. At Musk's request, the Autopilot team had designed and recorded a 'demonstration of the system's capabilities,' Elluswamy said. But there were problems. During testing, the driver had to intervene several times to prevent crashes. As a precaution, Tesla filmed the drive along a pre-mapped route. A parking scene – where the car hit a fence – was cut from the video.

Nonetheless, Tesla published the polished version on its website. Elluswamy considered that justifiable. 'The intent of the video was not to accurately portray what was available for customers in 2016,' he explained. 'It was to portray what was possible to build into the system.'

We start building a timeline. We want to track exactly when Elon Musk made which promise. The documents from the Tesla Files read as even more explosive when viewed in this light.

Take, for example, an Excel spreadsheet titled *NO Employee Vehicles false braking*, dated October 2020. 'NO' stands for Norway. Tesla employees in the Scandinavian country had documented how their vehicles, with Autopilot engaged, would brake on their own – without any clear reason, and sometimes at high speed.

Their verdict was damning. One noted that sudden braking occurred 'once or twice per drive.' Another reported experiencing phantom braking or abrupt slowdowns 'on almost every drive, usually a couple of times on my longer weekend drives at different locations.' Several employees admitted they had stopped using Autopilot altogether after such incidents.

None of them could identify a cause. They shared their theories in a spreadsheet. One suspected that his Model 3's Autopilot was misinterpreting speed limits on road signs. In any case, his car braked 'multiple times per day' for no apparent reason. Oncoming traffic on narrow roads also seemed to confuse the system.

Another employee noted that his car would always slow down at the same spot, which was just before a tunnel. He tested the section with ten different Teslas and wrote: 'Every car I've tested slows down there – for no reason.' A third speculated that 'deep shadows under bridges might also have something to do with it.' His Tesla had braked hard three times in three weeks – each time while overtaking on the highway.

Built to Fail

Regulators are finally taking action. Six months after our first report raised doubts about Tesla's Autopilot, the company is forced to recall more than 2 million vehicles due to accidents linked to the software. The U.S. National Highway Traffic Safety Administration (NHTSA) cites a 'critical difference between drivers' expectations' of Tesla's technology 'and the system's actual capabilities.'

The NHTSA oversees road safety. It has monitored Tesla for years. The agency originally approved Autopilot — but only on the condition that drivers constantly monitor the road and be ready to intervene if necessary.

In its recall notice, Tesla acknowledges that Autopilot's safety checks 'may not be sufficient to prevent driver misuse' and could increase the risk of an accident. The company pushes out an over-the-air software update with added driver alerts designed to ensure hands remain on the wheel. But the NHTSA isn't satisfied. At the end of April 2024, it is revealed that the agency is now investigating whether the new safety features go far enough. The concern: even updated vehicles were involved in crashes in which Autopilot played a role.

That same day, the NHTSA closes a three-year investigation into Autopilot. The agency reviewed 953 reported accidents between January 2018 and August

2023. In 489 cases, there was insufficient information for a full investigation. In 211 crashes, the Tesla hit another vehicle or object. According to the data, Tesla drivers in 59 of those accidents had at least five seconds to react before impact – 19 of them had ten seconds or more. The NHTSA concluded that the drivers involved 'were not sufficiently engaged in the driving task,' and that Autopilot's warnings 'did not adequately ensure the driver remained attentive.'

In May 2024, it's reported that the U.S. Securities and Exchange Commission is investigating whether Tesla violated securities laws in connection with its driver-assistance systems, Autopilot and Full Self-Driving. Reuters reports the SEC is reviewing whether Tesla misled investors or customers in its public statements. The U.S. Department of Justice is also examining whether Tesla or its executives made deceptive claims about the technology's capabilities.

Meanwhile Tesla has begun to soften its language – if only cosmetically. For years, the company's manuals have stated that drivers must keep their hands on the wheel at all times. But only since September 2024 has Tesla labeled the Autopilot option on its website's configurator as 'Full Self-Driving (Supervised).' The contradiction in terms doesn't seem to bother Tesla – as long as it offers some protection from legal challenges. In its latest software-update notes, the company states: 'Under your supervision, FSD (Supervised) can drive your Tesla almost anywhere.' But the system, it clarifies, does not make the vehicle autonomous. Hence the warning: 'Do not become complacent.'

The fact that Tesla was allowed to market its product for so long without this disclaimer is one thing. The fact that the new label effectively functions as a legal shield is another. By requiring the driver to remain ready at all times, Tesla puts ultimate liability back in the driver's hands. In practice, this often means that when an Autopilot crash occurs, authorities investigate the driver – but not Tesla. Our reporting shows that even in fatal crashes with unclear causes, prosecutors sometimes fail to ask whether Autopilot was truly capable of what it claimed. Whether a malfunction could have played a role. After all, the logic goes, the driver should have intervened.

More and more Tesla drivers appear to be realizing they may have fallen for a marketing trick. Our Autopilot coverage has made us a point of contact for frustrated owners. Again and again, drivers and their lawyers reach out to share details of their lawsuits against Tesla. In one case, the Berlin Regional Court appointed an expert witness in a dispute over a Model 3. The expert concluded that Autopilot poses a danger on German roads. During test drives, the vehicle steered 'left toward oncoming traffic.' It responded 'relatively late to lane departures, prompting the driver to take over out of fear.'

A telling ruling came in spring 2022 from the Regional Court in Darmstadt. It ordered Tesla to take back a Model S and refund the full purchase price of €68,730. In its judgment, the court found that the software package 'Full Self-Driving Capability' was defective. The feature for recognizing traffic lights and stop signs, it said, was incompatible with the onboard computer and 'indisputedly unusable.'

In summer 2022, the Munich Regional Court ordered Tesla to reimburse most of the €112,000 purchase price of a Model X. The customer had reported phantom braking and argued that the system frequently failed to detect obstacles and braked unnecessarily. Tesla countered that Autopilot was not designed for urban driving. The court rejected that defense, stating that it was unreasonable to expect an average customer to manually activate and deactivate the system via the central display depending on whether they were on a highway, country road, or city street. If Autopilot remained active, the court said, it posed 'a significant hazard in urban environments to the driver and surrounding traffic.'

One headline-grabbing case involved a Belgian IT consultant who sued for a refund on his Model S. He too had reported phantom braking. In June 2023 – a month after our first Autopilot article appeared – an appeals court in Antwerp ruled in his favor. There was no doubt, the court found, that the vehicle braked unintentionally. The judges concluded that the Tesla had 'serious deficiencies in driving comfort and safety.' They ordered the company to repay the €158,600 purchase price, along with €25,000 in legal and expert witness fees. Their ruling stated: 'The fact that the buyer was repeatedly confronted with minor, moderate, and serious defects from the time of delivery is sufficient to establish that the vehicle is not fit for its intended use.'

As we write this book, new whistleblowers continue to contact us about Autopilot. In September 2024, Tesla stands trial in Traunstein Regional Court over phantom braking in a Model 3. The customer had ordered the car in March 2022 for €60,920 and received it in December. He claimed that the vehicle repeatedly braked for no reason – especially near tunnels, large vehicles, or when the road surface changed. After Tesla failed to resolve the issue, the customer took the case to court in March 2023, seeking a replacement vehicle. Because Tesla's lawyers denied the phantom braking occurred, the court appointed an independent expert.

Starting August 6, the expert drove the customer's car roughly 700 kilometers, mostly along the A3 and A9 motorways between Munich, Ingolstadt, Nuremberg, and Würzburg. Two cameras recorded the drives from inside the car. The expert documented 'abnormal braking behavior' in five situations. In four of them, he was able to override the system. Twice, the car slowed in construction zones due to the system misjudging the distance to vehicles in the next lane.

The expert felt compelled to end the test drive after the fifth incident. Video we reviewed shows him driving in the left lane at 140 km/h on a three-lane highway when the car suddenly brakes to 96 km/h. There is no apparent reason for the maneuver – no signage, no traffic conditions. The expert says he narrowly avoided a crash. He wrote: 'The incident created significant danger for following traffic. Evasive maneuvers and hard braking were observed among vehicles behind.' He concluded that continuing the test drive 'on public roads without closed-off highway sections' was 'no longer safe to continue due to repeated braking failures.'

The Policy of Silence

The stories customers tell us often follow a familiar pattern: Tesla appears determined to avoid written communication altogether. Despite months of issues and repeated complaints, many say they've never received a single reply in writing. Almost everything is handled verbally. In the U.S., this is called avoiding a 'paper trail' – something Tesla appears to do by design.

We search the Tesla Files for internal guidance on how the company handles complaints like these. Sure enough, the data includes detailed instructions for employee–customer communication. Rule number one: leave as little of a trace as possible.

The files include screenshots from Tesla's internal system, documenting cases like that of customer Thomas Karl. When employees investigate an issue, they enter technical notes in the system. But the customer never gets to see them. The reports are consistently marked 'for internal use only.'

In every case, one directive stands out in bold: 'If you must share information with the customer, do so VERBALLY. Do not copy the Engineering Review into an email, text, or voicemail.' Furthermore: 'DO NOT provide any vehicle log data to the customer without explicit permission.' If, despite these efforts, legal involvement becomes inevitable, that fact must be logged immediately.

Elon Musk sees himself as the head of a dream factory. 'We all strive to provide an exceptional customer experience,' reads one of his internal guidelines on how to handle angry Tesla owners. 'We understand that for most of our customers, they are not just getting a new car, they are achieving their dream.' Sometimes, Musk admits, 'it's hard to fulfill a dream without some hiccups. When customers have a poor experience, they may ask to speak to a manager to resolve it.' Tesla's advice to its staff in such situations? 'Do not panic, put yourself in the customer's shoes and help them feel heard.'

Throughout our investigation, we often wonder whether Musk truly grasps what he's asking of his staff. Should they really put themselves in the shoes of the mother whose Tesla suddenly accelerates in the school parking lot? Or the driver whose car brakes in the middle of a tunnel? The countless pleas for help reveal how fast the Tesla dream can become a nightmare.

But in Musk's world of marketing myths, none of that seems to matter. In October 2024, he makes yet another promise. At a Hollywood studio in California, Musk unveils Tesla's first self-driving taxi: the Cybercab. The presentation is vague – no dates, no production details. But Musk is emphatic about one thing: 'We think that we'll be able to have driverless Teslas doing paid rides next year.'

CHAPTER 5

Promises Broken

In October 2024, a California court dismisses a lawsuit accusing Elon Musk of misleading investors about Tesla's Autopilot. While shareholders view his never-ending, never-fulfilled promises about self-driving cars as a deliberate effort to inflate Tesla's stock price, the judge sees no harm, no foul.

Musk celebrates the ruling as a personal triumph. On his platform X, he shares a post about the verdict and adds his own comment: 'Justice prevails.' His fans share the post more than 12,000 times; nearly 100,000 mark it with a heart.

To his critics, the decision is a triumph of misdirection. It wasn't Tesla's technology that prevailed – it was a legal tactic that painted Musk as a harmless exaggerator and suggested that only fools would take him seriously. As the company's lawyers argued: 'Vague, generalized assertions of corporate optimism or statements of "mere puffing" are not actionable material misrepresentations under federal securities laws, as no reasonable investor would rely on such statements.'

There is a name for this kind of legal strategy: the Puffery Defense. The term dates back to a case from 1892, when a customer sued the English company Carbolic Smoke Ball. The firm sold rubber balloons that emitted carbolic acid fumes, advertising them as protection against Russian flu. They even promised £100 to any customer who fell ill despite using the product.

When someone demanded that payment, Carbolic refused. In court, the company argued that their statements were 'mere puff' – marketing exaggeration, never meant to be taken literally. Carbolic lost the case, but it left behind an idea that has echoed ever since: advertising promises are not always meant to be believed. One hundred and thirty-two years later, that very logic forms the cornerstone of Tesla's legal defense – and of Elon Musk's public persona.

And there was plenty to defend. Musk's declarations such as Full Self-Driving being 'on the right track' and 'expected to be available by the end of the year' were only a few among many. He called Tesla's cars 'absurdly safe,' described the Autopilot as 'superhuman,' and said the system was 'as close to perfection as possible.'

Well, no. In court, it wasn't journalists or critics who undermined Musk's credibility – it was his own lawyers.

All those promises, all that bravado? Mere puff. Not to be trusted. No one should believe this man, they argued. Wasn't that obvious?

Yet no matter how unflattering a portrait Musk's legal team painted of their client, they achieved what they came for: Tesla was not held accountable for what Elon Musk promised.

For us business journalists, Tesla's Puffery Defence feels like an alien encounter with corporate logic. We know, from long experience, how obsessively companies control their messaging. Talented young managers, so-called high potentials, are given media training. *How do I behave on camera? How do I avoid saying something that might be used against me?*

Every CEO dreads what once happened to Rolf Breuer. In February 2002, the Deutsche Bank executive gave an interview to Bloomberg. Asked about the German media group Kirch, he said: 'Everything one reads and hears about it suggests that the financial sector is no longer prepared to provide further debt or even equity capital on an unchanged basis.' Two months later, Kirch filed for bankruptcy, triggering one of the most spectacular corporate collapses in postwar German history. Kirch's founder blamed Breuer directly: 'It was Rolf who pulled the trigger.'

What followed was a decade-long legal battle. In the end, Deutsche Bank paid nearly a billion euros to Kirch's heirs. The Breuer interview now has its own Wikipedia entry. It's considered a textbook case of how a single careless statement from a CEO can trigger disaster.

Every year, countless seminars are held on the importance of so-called CEO communication. 'The CEO is the projection surface for success or failure; he is the face of the company,' states a corporate communications handbook. 'CEO communication is the strategic shaping of the public perception of the CEO in order to advance the company's business agenda.'

Any business journalist can recall the near-paranoia within corporate comms departments: the fear that even a single ambiguous sentence might slip through. As the investigative team at *Handelsblatt*, we know that answers to our inquiries are often filtered – first by PR staff, then legal teams, sometimes multiple ones.

At Tesla, it's a different universe. Musk dismantled the communications department. He dismisses critical reporting as fake news or hit pieces – journalistic firing squads. Sometimes he sues. But when someone tries to sue him for the reckless, empty promises he makes, Musk reaches for the dumbest defense of all: *Why should I care about what I said yesterday?*

The California case shows that Musk succeeds with this approach. Like a conductor guiding an orchestra, he plays with the fantasies of his fans and shareholders with style and theatrical precision. His career is built on making promises about the future. These predictions have made him and many of his investors unimaginably rich.

They've also helped craft his image as a visionary. The media, always hungry for hero stories, treated every new Musk announcement as gospel – often for years, without ever circling back to ask: Did any of this actually come true? That has begun to change. Public skepticism toward Big Tech's salvation promises has grown. Musk's bold claims have become memes, punchlines, viral jokes. More and more often, under posts about failed products or dashed hopes, you'll find the same wry comment:

'You got Musked.'

Few companies show a wider gap between image and reality than Tesla. Sure, Musk – or a successor – might someday make parts of his vision come true. But the long trail of broken promises raises a sharper question:

Is Elon Musk a better storyteller than an inventor?

And how many more hollow promises will it take before people stop mistaking his fantasy for fact?

The Price of Hype

The dispute over Musk's Autopilot wasn't the first time the Tesla CEO turned to the Puffery Defense. He also invoked it in a lawsuit over his statements about the cryptocurrency Dogecoin, arguing that no one should take him seriously.

Musk had repeatedly expressed his enthusiasm for the coin in various tweets, significantly boosting its value. Among other things, Musk suggested that Dogecoin could be the 'future currency of Earth.' He said he believed Tesla would eventually accept payments in the cryptocurrency. At one point, he even replaced the Twitter logo with the Dogecoin symbol. Musk's stunts earned him the nickname 'The Dogefather'.

Musk once explained his behavior to employees at Tesla's German plant in Grünheide by saying that workers at the factory had asked him to do it. People working on the assembly line had asked him, '"Can you please support Doge?",' Musk said in a speech. At SpaceX, too, he claimed, 'a couple of regular guys' had asked him the same thing. So Musk decided: 'You know what? Doge is people's crypto, so I will support it. Because lots of rich people were supporting Bitcoin, but I was like, "Oh, but people on the line want me to support Doge, I'll support Doge."'

In June 2022, Dogecoin buyers filed a lawsuit against Musk, seeking $258 billion in damages. They argued that Musk had artificially inflated the price of Dogecoin by 36,000 percent through his tweets, leading investors to make unwise decisions. According to the suit, many investors suffered major losses when the coin's value later plummeted.

Musk's influence on the price of the cryptocurrency, they claimed, amounted to market manipulation. After his glowing praise, Musk had referred to Dogecoin on television as 'a hustle.' Only Musk knew whether this latest remark would send the coin soaring or crashing.

In the legal battle over these remarks, Musk's attorneys used the same argument as in the Autopilot case: Musk's announcements and praise for Dogecoin, they explained, were 'innocuous and often silly tweets.' They prevailed – the lawsuit in New York was dismissed at the end of August 2024. Once again, the Tesla CEO was off the hook.

Is it too early to call this a pattern? Too late? People believe Musk's insane promises because he is the richest man alive. What they overlook is that those very promises are what made him the richest man in the world in the first place. And so, the Muskian cycle completes itself.

His most spectacular project is SpaceX. As we write this book, Musk is achieving a historic breakthrough with his space company. After a pioneering test flight reaching speeds of over 27,000 km/h, the booster of the Starship rocket is returning safely to the launch site in Texas. Musk hails the mission as a key step toward his vision of extending human life to other planets.

While boasting about being a pioneer, Musk is far behind his own schedule. In 2014, he said in an interview: 'I'm hopeful that the first people could be taken to Mars in ten to twelve years.' In February 2017, Musk announced that by the following year, he would enable two space tourists to circle the moon. Shortly thereafter, he shared that

SpaceX rockets would one day carry passengers from New York to London in 29 minutes and for the price of an economy-class plane ticket. None of it came true.

Musk's space promises echo his endlessly repeated predictions about autonomous driving. Meanwhile, Autopilot isn't the only product with which Musk wanted to revolutionize how we move. On December 17, 2016, the entrepreneur tweeted from his Tesla while stuck in gridlock in Los Angeles: 'The traffic is driving me crazy. Am going to build a tunnel boring machine and start digging.'

That same day, he founded the Boring Company. Musk announced that his latest startup would build a tunnel-based, automated transport system for cars to solve global traffic problems. He had published a white paper laying out his concept for this three years earlier. Musk wrote of a 'Hyperloop' – a system of electric sleds capable of reaching speeds of more than 1,000 km/h.

At the end of April 2017, the Boring Company released a video on YouTube. It showed a Tesla navigating through Los Angeles traffic. The vehicle then pulled onto a roadside platform that descended into the ground like an elevator. Once inside the tunnel, the platform – with the car on it – merged into underground traffic and sped the Tesla through the city at up to 200 km/h. At the destination, the vehicle and its passengers rose back to the surface and rejoined regular traffic.

On July 20, 2017, Musk tweeted that the Boring Company had received verbal government approval to build an underground Hyperloop connecting New York City, Philadelphia, Baltimore, and Washington, D.C. The system would allow for travel from New York to Washington in just 29 minutes – echoing his SpaceX flight claim from London to New York. It later turned out that all Musk had received was an informal offer of support from Jared Kushner, the son-in-law of then-president Donald Trump. Four years later, the Boring Company quietly deleted all references to this and other supposed projects from its website.

In 2018, Musk announced at a press conference with the mayor of Chicago that the Boring Company would build a tunnel in which passengers could travel the 29 kilometers from downtown Chicago to O'Hare Airport in just 12 minutes. A new promotional video showed autonomous vehicles, each seating 16 people, zipping through underground tunnels.

This project, too, went nowhere. In May 2018, Musk declared that the Boring Company would use earth from its tunneling to build affordable housing. As of today, the company has completed only a single project – and did not use any excavated earth for construction. In the Boring tunnel in Las Vegas, there are no autonomous 16-passenger vehicles, nor is there high-speed transportation. Underneath the Convention Center, it's just ordinary Teslas, driven by people.

Cyberstuck

In the California Autopilot trial of late 2024, a key point in Musk's defense was that the plaintiffs couldn't prove he had known better. According to the judge, the fact that Musk's promises hadn't materialized wasn't sufficient. The plaintiffs would have needed to show that the Tesla CEO had knowingly exaggerated. As long as he claimed to believe in his own visions, he could not be held liable for their failure.

For Musk, this may seem like a license to keep hyping himself and his companies with impunity. However, there's at least one case that casts serious doubt on Musk's presumed good faith. It concerns the very product he spent years portraying as Tesla's great beacon of hope: the Cybertruck.

Tesla aimed to use the bulky electric pickup to break into the multi-billion-dollar U.S. truck market. In July 2016, Musk called such a vehicle part of his 'master plan.' In 2018, he tweeted: 'I'm dying to make a pickup truck so bad . . . we might have a prototype to unveil next year.'

On November 21, 2019, the Tesla CEO stood on stage in Los Angeles with the new Cybertruck beside him.

Musk emphasized the vehicle's toughness: a bulletproof shell, windows made of unbreakable armored glass. Then came the counterevidence. At Musk's request, Tesla's chief designer Franz von Holzhausen hurled a metal ball at one of the Cybertruck's windows. It shattered instantly. 'Oh my fucking god,' Musk blurted out. 'Well, maybe that was a little too hard.' He then suggested Holzhausen try a second window. That one shattered too.

Musk made new promises – about the price, the range, and the truck's capabilities. 'If we get lucky, we'll be able to do a few deliveries toward the end of this year, but I expect volume productions to begin in 2022,' he said in January 2021. 'We finished almost all of the Cybertruck engineering.'

A year later, production was still a mere fantasy. In Texas, Tesla's engineers put together a slide deck titled 'Cybertruck Ride & Handling / NVH.' NVH is industry shorthand for noise, vibration, and harshness.

The PowerPoint document is part of the Tesla Files. When we first open it, we can hardly believe our eyes. The document is deeply technical, and since none of us in the investigative team studied automotive design, we have trouble making head or tail of it. Yet even so, one detail stands out immediately: the report uses a traffic light system. Goals met are marked in green, missed ones in yellow, and major failures in red. The tables are awash in yellow and red. We don't need a PhD to know something is not right with the latest dream machine of Elon Musk.

We study the document – 20 pages of test data and performance metrics. We delve into the engineers' methodology. We consult with industry experts. In face-to-face meetings, we show selected excerpts from the report. Most are stunned by the range of issues Tesla is battling. Andy Palmer, former Nissan executive and ex-CEO of Aston Martin, calls the presentation 'troubling.' It points to one conclusion: Tesla's hope for billions in profit is structurally compromised and scores poorly in nearly every category drivers care about – steering, acceleration, and braking.

Test drivers reported loud cabin noise under certain conditions. The pickup jerked during parking. The air-conditioning system apparently caused the steering wheel to vibrate. At high speeds and during braking, the proto-type lost stability. What hits us harder than the test results is the date on the file: January 25, 2022. According to internal records, Musk had already reviewed it six days earlier. The Tesla CEO was fully conscious of how far behind schedule the Cybertruck had fallen, how long it would be till this new car had any meaningful impact on his balance sheet.

And yet, here was Musk on January 26, tweeting: 'Been driving latest Cybertruck prototype around Giga Texas. It's awesome!'

It still isn't. Tesla delivered the first production Cyber-trucks in November 2023 – nearly two years behind schedule. The truck was more expensive than Musk had promised, had less range than claimed, and – contrary to earlier hype – was not 'waterproof enough to serve briefly as a boat, so it can cross rivers, lakes & even seas that aren't too choppy.' Musk had pitched the Cybertruck as apocalypse-ready and fit for any planet. When videos of Cybertrucks stuck in the snow went viral, the internet gave Musk's passion project a new name: Cyberstuck.

What's left after all the hype? Elon Musk claimed over a million preorders, and Tesla initially vowed to produce more than 100,000 Cybertrucks in 2024, scale up rapidly, and sell 250,000 units within 18 months. Yet data compiled by the market research firm Cox Automotive tell a different story: just 39,000 Americans bought a Cybertruck in 2024, with sales in the third quarter as the high point and fourth-quarter deliveries down 22 percent. If year one disappointed, year two has been catastrophic: only 6,400 delivered in the first quarter of 2025 – a 50 percent drop from the fourth quarter of 2024. By May, even $10,000 discounts haven't moved stock. Media reports suggest around 10,000 Cybertrucks remain unsold across Musk's empire. Hailed by its creator as Tesla's 'best product ever,' the Cybertruck has become a disaster of epic proportions.

Then there were the technical flaws. Tesla has already had to recall the Cybertruck seven times – for reasons varying from faulty windshield wipers to malfunctioning reversing cameras. The U.S. highway safety regulator stepped in after customers reported that the aluminium pedal cover had come loose and jammed between the pedal and the footwell.

In China and Europe, the Cybertruck is on display only. Musk has confirmed that it needs to be redesigned to meet regulatory standards. That, he said, would only make sense once production volumes increase.

In a March 2024 talk with employees in Germany, Musk offers little hope that the electric pickup truck will be coming to Europe anytime soon. 'So, Cybertruck does not match the EU rules, which are much more strict than the U.S.,' he says. 'But I think we might be able to do, at some point, a little bit smaller and it would look a little bit different, but it does match EU regulations. So, maybe we can look at doing that in a few years to try to bring sort of more of an international version.' For now, Musk urges, the focus has to be on 'ironing out the production challenges of a totally new technology.'

At Tesla's German Gigafactory, the company runs a weekly raffle for a ride in the U.S. version of the Cybertruck. Each week, employees receive a quiz on factory-related topics. Those who answer correctly are entered into the draw. Winners get to drive a lap around the rear part of the factory grounds in a Cybertruck. Management says over 1,000 people – sometimes more than 2,000 – participate each week.

With those numbers, the promotion is a hit. Until it isn't. As employees tell us, the quiz Cybertruck has already broken down twice. To some, it's already a symbol – not of the future, but of a promise stalled in the parking lot.

Launch Now, Build Later

Tesla employees in Germany are especially familiar with Elon Musk's empty promises. In November 2023, the CEO took the stage inside a large white tent on the Berlin factory grounds. Musk arrived in jeans and a bomber jacket, as shown in a video shared with us by his employees. He thanked them for 'the excellent, hard work in creating what is, I think, one of the greatest factories in the world.' With the German site, he said, Tesla had created 'a real crown jewel. And it looks like a crown jewel. It's beautiful.'

Then he talked about upcoming products and made a surprising announcement: 'We're going to have a super-high-volume, smaller car. I can't say too much about it right now, but obviously we want to make that here. It's going to be really fantastic.'

Musk was referring to the mass-market vehicle he has been promising for years – target price: under €25,000. Originally, this still-unnamed model was to be manufactured in Texas or the planned factory in Mexico. Now Musk was offering his German Giga team work for years to come. 'Like I said, the thing that's really going to revolutionize the future is an ultra-high-volume, super-affordable car. It's still a great car, but one that's made, you know, it's several times the volume of our current vehicle. That is affordable to basically anyone. But it's still a great car. That's really the future.'

When Musk returned to the factory four months later, he repeated the promise. Asked by an employee about the budget model, Musk said, 'the high-volume vehicle will definitely come to Berlin long-term, yes.' And there was more. 'We're aiming to make a semi-truck where the price of the semi-truck is roughly similar to the price of a diesel semi-truck. That's our goal,' Musk said. 'And I think it would make sense to have the semi-truck production also in Europe at Berlin.'

For Giga Germany, the announcement was massive. Tesla was the biggest employer in the region, and now it was going to get even bigger. The cheap Tesla. The Semi. 12,000 people already worked at the plant; now even more jobs were coming. Unless . . . could Musk have been puffing again?

Yes. In October 2024, Musk had nothing but disdain for his own words. Asked by a financial analyst when investors might expect the affordable model, he said: 'So, anyway, basically, I think having a regular 25k model is pointless. It would be silly. Like it would be completely at odds with what we believe.'

To date, neither the Semi nor the budget model is being built in Germany. As of January 2025, the factory is rolling out a revised Model Y. Otherwise, it's the same old Teslas, while workers are still being told by management to do their part to help fix the company's sluggish sales. More than that: Tesla has put its planned factory expansion on hold. In August 2024, plant manager André Thierig said, 'We're not going to spend billions on expanding the plant unless it's absolutely clear that the market is asking for it.'

The message seems clear: Whatever Musk says, even his own employees would be fools to believe him. And yet, Musk wouldn't be Musk if he didn't have another card up his sleeve. Alongside denial, he's a master of bait and switch. No sooner had he scrapped the budget model than he dangled a new carrot in front of employees and shareholders: the robotaxi.

This one is a Musk classic. In 2019, he declared that Tesla would 'have a million robotaxis on the road' by the following year. In April 2020, the actual number was still zero, but Musk chided anyone who doubted him, telling analysts on an earnings call: 'Punctuality is not my strong suit, but I always come through in the end.'

Musk reiterated: 'I think we could see robotaxis in operation . . . next year. Not in all markets, but in some.' 2021 came and went without a single robotaxi. In April 2022, speaking again to analysts, Musk said: 'I think can be very powerful product where we aspire to reach volume production in 2024.' He added: 'I think [the robotaxi] really will be a massive driver of Tesla's growth.'

Once operational, the robotaxi fleet would give people the Tesla experience even if they could never afford to own one, Musk promised. 'Look at some of our projections: It would appear that a robotaxi ride will cost less than a bus ticket, a subsidized bus ticket or a subsidized subway ticket.'

In July 2024, Musk announced the product launch. Shareholders across the globe got excited. Over the next few weeks, Tesla's stock jumped more than 40 percent. The robotaxi was first set to debut on August 8, but a last-minute design tweak by Musk pushed the reveal to October 10. Two weeks before that, he posted on X: 'This will be one for the history books.'

And then, it happened. Same idea, new name. At the 'We, Robot' event held at the Warner Brothers Studios in California, Tesla unveiled the Cybercab: a futuristic, silver, sci-fi-looking two-seater – no steering wheel, no pedals, gullwing doors.

Dressed all in black, from leather jacket to shoes, Musk arrived at the venue in a Cybercab and radiated pride. Only, when he stepped out, he had almost nothing of substance to say. He spoke about not wanting to live in the kind of bleak future portrayed in many science fiction films. He said he wanted to live in a future that was fun. He talked about traffic in Los Angeles.

Musk played all the hits. How cars were underused and could do a lot more if autonomous. How soon people could get into a Tesla, fall asleep on the back seat and wake up at their destination. How Autopilot was going to drive cars ten times more safely than any human could. Then, with hardly any information as to how, when and where his new vehicles were going to be built, Musk declared: 'The autonomous future is here.'

After months of delay, the 'historic' reveal started almost an hour late and ended in under 30 minutes. On Wall Street, it bombed. The day after the event, Tesla's stock fell nearly 9 percent – a $67 billion loss in market value. 'We found Tesla's robotaxi event disappointing and notably lacking in detail,' said analyst Toni Sacconaghi. Industry expert John Colantuoni called the presentation 'toothless' and predicted that competitors would now benefit – like Uber, which had just announced a partnership with Waymo, Google's self-driving car division. Sure enough, Uber shares rose 9 percent after Tesla's event. Rival Lyft was up 10 percent.

Musk had one more promise to offer that day. Toward the end of the event, humanoid robots marched into the room. They, too – named Optimus – had been promised years earlier. 'We've made a lot of progress on Optimus,' Musk said. The robots would now mingle with guests. 'They're also going to serve you drinks at the bar.' Then the robots danced on stage to show off their mobility.

It was the first time that non-employees had a chance to interact with Optimus in person. They gave them high fives and played rock-paper-scissors. Musk said Optimus could be a teacher, take care of children, walk the dog, mow the lawn, do the shopping, just be a friend, serve drinks – 'it can basically do anything you want.'

As for timelines or revenue projections, Musk said nothing. Anyone who asked got a curious answer. 'It is our understanding that these robots were not operating entirely autonomously – but relied on tele-ops (human intervention),' a Morgan Stanley analyst said after the event. Bloomberg reported that the robots had been partly remote-controlled to impress investors. Walking had apparently been no problem, but other actions required staff assistance. One robot is even said to have admitted, when asked, that it was being controlled by a human.

Musk, minutes earlier, had made no mention of this – only grand claims about 'the greatest product ever.' He added that Tesla's path to autonomy and AI could push the company's valuation to $5 trillion – around seven times its current worth (October 2024).

Musk's playbook is now well known. He is selling tomorrow, today. Always overpromising, always under-delivering. What Musk runs isn't just a portfolio comprising a car company, a space firm, a tunnelling enterprise and more – it's a conglomerate built on bold projections and glossy presentations. The tech world has a name for that: vaporware.

It refers to products that are heavily promoted but either never materialize or arrive years behind schedule, often in a form that falls far short of their original claim. The term, coined in the 1980s, combines 'vapor' – as in, insubstantial or elusive – with 'software' or 'hardware', and has since come to describe tech entrepreneurs' habit of selling dreams long before delivering reality.

Elon Musk is the worst of them. 'My mind often feels like a very wild storm. I have a fountain of ideas. I have more ideas than I could possibly execute, so I have no shortage of ideas,' Musk said in an interview with the *New York Times*. 'Innovation is not the problem. Execution is the problem.' But as long as being innovative is as valuable as actually executing those ideas, why mess with the playbook?

There's every reason to believe that Musk will keep making promises. If there's one thing he has, it's imagination. As several insiders told us, Musk had already been talking to employees in his German Gigafactory in Grünheide about a new project. For the next version of Tesla's Roadster sports car, he said, Tesla would partner with his space company, SpaceX. Musk, for his part, was already hooked. The new Roadster would be 'the best sports car ever' and 'do a few crazy things no car has ever done before. So it'll be very cool. Very cool indeed.'

That promise isn't new either. Musk has previously said SpaceX might develop propulsion systems for Tesla. The new Roadster, he once suggested, might even fly. In Grünheide, Musk told workers: 'The Tesla Roadster will be kind of a rocket car.'

Does that sound insane?

Maybe.

But across Musk's empire, the pattern rarely changes: a spectacular promise, a wave of headlines, a stock bump – and then reality catches up. Autopilot isn't autonomous. The Cybertruck isn't indestructible. Dogecoin is a joke. The Hyperloop isn't high-speed. And the robotaxi? Still a PowerPoint. What matters isn't delivery – it's belief. Musk's product is the promise.

What does he have to lose? If shareholders believe him, then someday feel burned, his legal team will be ready, with its favorite line: Musk wasn't lying. He was just puffing.

And if we fall for it – well, that's on us.

CHAPTER 6

The Whistleblower

My [Sönke Iwersen's] smartphone seems unusually quiet today. I've just tucked my little daughter into bed and read *The Robber Hotzenplotz* – a classic German children's tale about a mischievous bandit – to my son. I'm sitting on the sofa with a cup of tea when it strikes me. Over the past few weeks, I've exchanged dozens of messages with Lukasz – the Tesla whistleblower – every single day. Today, silence. Is something wrong?

It's been six days since we published the Tesla Files. Now, our reporting on Elon Musk and his business practices is in the international spotlight. Hundreds of articles have been published – from London and New York to Beijing. Major U.S. newspapers are asking to look at the data we obtained. On Saturday, during my son's soccer match, *Wired* called and proposed a collaboration. The biggest labor union in

Germany, IG Metall, wants to talk to us. Several authorities have launched investigations.

Lukasz has spent the past few days in a kind of trance. The global media echo, the mounting investigations, the constant stream of social media posts – it all electrifies him. At the same time, one thought keeps gnawing at him: he's taken on Elon Musk, the richest man in the world. What will happen to Lukasz now?

Caution is advised. 'I should probably take more serious steps about my privacy,' Lukasz messages me the day before, just before the evening news. 'New number. Maybe new phone. Reinstall the systems on computers. I am sure T put a lot resources into forensics now.'

We've been communicating over an encrypted messaging service for half a year now. Referring to Tesla as 'T' has become a fond habit. At first a mix of awe and worry, shortening Tesla to 'T' is now a bit of a nod to ourselves. We did it, right? After months of work, the article based on Lukasz's Tesla Files is finally published. Yet, caution seems like the sensible course of action. I encourage his idea of being extra careful.

'Good idea,' I reply. But it's too late, as his first message today makes clear. 'Hi Sönke, FYI, Police searched my house with bailiffs this morning and took all my electronic devices and accounts. Just got back the most essential devices.'

Oh hell. I pause for a second, feeling sorry for Lukasz. I've known him for half a year now. He's an emotional person, quick to anger. There were moments over the past months when he seemed depressed. He's unemployed, living alone in a foreign city. And now this. Police. A raid.

I'm shaken too, and I sympathize. Then something kicks in – a kind of occupational reflex. A cold, journalistic mechanism starts running in my head. A raid on a whistle-blower is news. How do we verify it? If there was a search, there has to be a warrant. Can we get it?

I ask Lukasz: 'Did they give you a written explanation?'

He confirms. But he can't scan the document – his computer was confiscated. I ask: 'How are you? Did you have to go through this alone?' His answer: 'Yes, alone. No phone no access to anything. Could not reach my lawyer. I am better, but whole day I was dizzy, couldn't focus staring at the wall or floor for long periods of time with thoughts running about all events at T.'

We keep messaging, carefully. Both of us know someone might be reading along. Lukasz had to hand his phone over to the police. Who knows what apps are now running in the background? At 11 p.m., I tell my wife I need to make a call. I walk into the kitchen and dial Lukasz's number.

We've talked often about how long he can – or wants to – stay anonymous. After some initial reluctance, Lukasz has always said he knows his identity would eventually be revealed. Neither of us expected it would happen just six days after our first article. Who could have anticipated that the Norwegian police would act so swiftly on Tesla's request?

I ask Lukasz again how he's doing. His voice is shaky. 'I'm nervous,' he says. The day has taken its toll. 'I don't know what I can do.' Ever since he's been in a labor dispute with Tesla, Lukasz has a lawyer who specializes in this field. But now the legal accusations are of an entirely different nature. How is he supposed to find the right protection and advice so quickly?

Lukasz asks what I expect now. Will *Handelsblatt* be raided too? Is that something Tesla can facilitate as well? I try to reassure him. I've been a journalist for more than 20 years. Lawsuits are nothing new – but a raid? That's uncharted territory. Still, I mentally file the question away. We talk for a few more minutes. I tell him to try to get some sleep – despite everything. He'll need his strength.

Two minutes later I text my investigative colleague Michael. 'Are you still up?' I can't help but smile at the line – moments later I see the familiar three dots pulsing on the screen. Michael lives and breathes this story. Weekends, holidays, vacations – everything in the past six months has been soaked in Tesla. 'We should talk briefly,' I write. Then I call him. It's nearly midnight.

Michael says we know what to do if the worst happens. A year and a half ago, during a different investigation, we took part in a workshop on police searches. Back then, we had published a report that was classified as secret by the Bundestag, the federal parliament of Germany. The report included compromising information about the Wirecard scandal, a fiasco that cost investors billions and called into question the competence of not only Wirecard management, but also its auditors, Ernst & Young (EY).

EY filed a criminal complaint shortly thereafter. Since then, we've had Post-it notes sticking to our doors, displaying emergency numbers of lawyers to call in case of a surprise visit by German police.

Enough for today. I go to bed and sleep well. Sometimes colleagues ask me if I can even relax during stories like Tesla. But stress has never been a problem for me. In the middle of the night, my daughter wakes up and protests. But a few minutes after I lift her from her bed, we're fast asleep beside each other.

The next day begins like always. Get up, wake the kids, check for leftover toothpaste on their faces. One change in the morning routine: between buttering bread and slicing apples, I call Peter Koppe, our in-house lawyer. Neither of us believes Tesla can convince the German police to raid a newsroom. But who knows. After the call, he texts me: 'I'll go ahead and put our criminal defense attorney on standby, just in case anything happens.'

Koppe recommends informing the whole Tesla Files team. Over the past few months, it's not just our investigative unit that has worked on the topic, but also the automotive team, U.S. correspondents, and others. We communicate through an encrypted group chat.

I write: 'ATTENTION. DON'T TELL ANYONE OUTSIDE THE GROUP: The source was searched yesterday. All electronic devices confiscated. We've spoken. He's doing okay, all things considered. Koppe is informed. He's alerting our criminal defense lawyers.'

The Raid

Twenty-four hours earlier. Lukasz grumbles as the doorbell rings. Still dazed, he reaches for his phone, opens one bleary eye, taps the screen. Seven a.m. What the hell is going on? He rolls over and pulls the blanket over his head. Please, not now.

The ringing turns into pounding. Lukasz forces himself out of bed, still groggy. Halfway to the door, he hesitates, moves to the window, peers out. A car is parked in front of the building – one he doesn't recognize. Carefully, he slips over to another window to get a view of the entrance. His stomach drops. Three officers in black uniforms stand outside. Lukasz begins to shake. Only one thought flashes through his mind: Tesla has found him.

The pounding doesn't stop. Lukasz feels his stomach tighten. Barefoot, wearing only a T-shirt and sweatpants, he steps to the door and cracks it open. 'Yes?' A woman stands before him, a stack of papers in her hand. She says something in Norwegian. Lukasz looks at her, confused.

He's lived in this country for five years, but his Norwegian is still shaky. The woman switches to English. 'Police. We have a search warrant,' she says, handing him the documents. Lukasz can't make out much, but one word leaps out at him from the front page: Tesla.

Another car pulls up outside. Two more strangers get out.

Lukasz hesitates. It won't be long before his neighbors wake up. What will they think of him? He knows these people – upstanding, deferential to authority. Anyone who has half a dozen cops at their door must have done something wrong. That's what they'll think. Lukasz opens the door fully and says, 'Come in.'

As soon as they step inside, the officer takes his iPhone. Procedure, she explains. She can't risk him erasing any data while her colleagues search his apartment. And just like that, Lukasz is cut off from the outside world.

Five strangers are now inside his home: three police officers, two employees from IBAS, a company specializing in digital forensics and data recovery. Helpless, Lukasz sits on his bed and watches as the authorities rummage through his two-room apartment. Every drawer is opened, every bag emptied. The search lasts four hours. When the officers leave, they take Lukasz's phone, his computer, his storage devices.

He's left behind, numb. For hours, he lies on his bed, arms crossed behind his head. He can't think straight.

Slowly, his mind begins to clear. At the end, he's left with one question: How on earth did I end up in this fight with Elon Musk?

The Disciple

Lukasz Krupski is born on November 3, 1985, in Warsaw. These are turbulent times. Four years earlier, head of state Wojciech Jaruzelski declared martial law. The cityscape is marked by long lines outside grocery stores; militia and security forces patrol the streets. Political graffiti covers many walls.

Lukasz's father is a student. When he takes a teaching job, the family moves from the big city back to his small hometown of Żaby, a village 180 kilometers southwest of Warsaw. The house Lukasz grows up in stands on his grandparents' land – a farm on a nameless road separating fields from forest. Before long, he has two sisters and a brother.

Lukasz is a strong student, curious about technology, especially anything that moves. As a boy, he climbs onto every machine he can find: tractors, combine harvesters, fertilizer spreaders. Later, he takes apart cars and puts them back together. Even as a teenager, he's a key helper on the farm.

But Lukasz doesn't want to be a farmer. He is drawn to aerospace. In 2004, he begins studying physics in Warsaw.

Then love intervenes. He meets a young woman who plans to study business management in Wrocław. Lukasz follows his heart – first to Wrocław, then into the lecture halls of the business university. When the relationship ends, he takes a break.

With a student visa, Lukasz moves to England and from there applies for a technician job in Canada. He earns money as an electrician in Denmark and on construction sites in Australia. For a Polish company, he conducts research on electric vehicles. That research takes him to a trade show in Munich in October 2011 – his first encounter with a Tesla.

That same year, Lukasz completes his business management degree. He returns to England and spends almost a year working on a farm. The owner, Nick How, is fascinated by Lukasz's instinctive command of machines and vehicles of every size. Even when he sees a machine for the first time, he quickly knows it inside out. But How also notices something else about his employee: Lukasz has a strong sense of justice – almost a tendency to rebel. There's a defiant streak in him, no doubt. How wonders what that trait might mean for Lukasz's future.

In the winter of 2017, Lukasz sees a job posting from Tesla. The chance to work for Elon Musk excites him immensely. A techie himself, Lukasz sees Tesla as more than just a company. To him, the American carmaker is a force for good. He had already applied for several jobs at Tesla, years earlier. When he receives an offer in the summer of 2018 to work as a service technician in Norway, he jumps at it, even though the terms are modest.

On October 2, 2018, Lukasz signs his Tesla contract. As a service technician, he earns less than in previous jobs. Tesla pays him an annual salary of 430,000 Norwegian kroner (€37,000).

When Lukasz puts on his employee badge for the first time, it feels like a dream come true. He's proud to work for Tesla. Proud to contribute to Musk's vision. He even calls himself a 'Musk disciple.' As employee number 709279, Lukasz is ready to help change the world.

Trial by Fire

Lukasz begins to question his decision to work at Tesla from his very first day. The company hasn't provided him with work clothes. When he picks up his laptop, coworkers tell him they waited months for theirs. Even basic tools – like a hammer or screwdriver – are missing. Lukasz finds himself wondering: *What kind of shit show is this?*

Yet Norway is a key market for Tesla. Nowhere in the world do electric vehicles make up a larger share of new registrations. No other country has more Teslas per capita. When Lukasz starts his job in Drammen, car carriers are bringing vehicles straight from the U.S. When they roll out of the containers, his job is to adapt them to local requirements.

Working at Tesla is no cakewalk. Most of Lukasz's co-workers – like him, immigrants from Poland and Lithuania – contribute to a rough environment. They curse, mock one another, and show porn clips on their phones. Anyone who doesn't play along quickly becomes a target.

The department culture spills over into a Facebook group called D2, named after Tesla's delivery center in Drammen. There, employees post sexist memes during work hours. One shows a flat-screen TV next to a woman. The caption: 'Look at the size of my new TV compared to my dishwasher!'

Daily work turns into an ordeal. Lukasz clashes repeatedly with colleagues. They see him as a nerd; he sees them as brutes. But all that fades into the background the moment a Tesla Model 3 goes up in flames. To this day, Lukasz says the smell of burning plastic still haunts him when he thinks of March 30, 2019.

He's still on probation when he risks his life for Tesla. That afternoon, the company is showcasing its new Model 3 in Lillestrøm. At 2:30 p.m., the unthinkable happens: a display car catches fire.

Lukasz is working just steps away when he feels a sudden heat on the back of his neck and smells something sharp and acrid. He turns and sees flames flickering from a car. It's a Model 3, parked close to a heavy curtain that separates the working area from the showroom. Just behind it, visitors are chatting and strolling around, completely unaware of the danger. Thinking fast, Lukasz rushes to the vehicle, reaches into the engine compartment, and pulls out a burning power booster, using only his shirt sleeve to shield his hand.

His quick action prevents disaster. Tesla vehicles run on lithium batteries, which even seasoned firefighters approach with caution; when water hits burning lithium, it can release highly flammable hydrogen – intensifying the blaze or triggering an explosion. Just seconds later, the flames would have reached the battery. Now the smoldering booster lies on the ground, and Lukasz's colleagues smother the last flames with blankets and floor mats.

Two days later, Elon Musk hears about the near-catastrophe. Tesla's head of service in Norway informs the CEO, praising Lukasz in an effusive email: 'I would like to send a special thanks to Lukasz Krupski, a service technician from Drammen, for his heroic effort on Saturday 30.03. With his bare hands he was able to pull it [the booster] out of the car before it was too late. This resulted in some severe burn damage to his fingers, but it is fair to say that if it wasn't for his action, the result would have been a car on fire.'

When the Norwegian boss suggests Musk send Lukasz

a token of appreciation, the Tesla CEO does so within hours. He emails Lukasz: 'Thank you and congratulations for saving the day!'

The man of the hour plays it down, replying to Musk: 'Thank you for your email and no worries really. It was the only reasonable thing to do at that moment. I had to do what was right.'

Musk follows up: 'Anything we can do to make things better in the future?'

He has no idea how much that question means to Lukasz.

The new Tesla technician sees risks everywhere. There are too few fire extinguishers. Cars are parked so close together that sparks and molten metal from grinding operations can hit nearby vehicles – or people. Some coworkers modify tools or components without understanding the consequences – even risking a disaster like the one in Lillestrøm.

On April 2, 2019, Lukasz emails Musk again. A technician, he writes, had caused the fire by tampering with a power booster and removing a safety device. Lukasz recommends banning those amplifiers entirely, adding a bit of philosophy: 'If something happens once it will very likely happen again. That's reasonable rule in this universe. Especially regarding to human nature.'

Musk appears grateful. 'Ok, please let me know if there's anything we should still do,' he replies on April 4. That same day, a Tesla safety manager sends an email to senior staff across Europe.

Subject line: 'SAFETY ALERT – Booster Battery Fire – PLEASE CASCADE.'

From now on, the message says, it is 'strictly forbidden to modify or change' any tools required to do a job. Referring to the fire in Lillestrøm, the manager warns:

'This will put you, your colleagues and the company at a huge risk!'

Again, Lukasz is praised. As his boss writes: Only the 'rapid response of the technician' had kept the incident from becoming far worse.

Lukasz is pleased. He believes Tesla's cars are making the world a better place. Now, six months into the job, he's helping Tesla become a better company. Even at the bottom of the hierarchy, his warnings are heard at the very top.

This fits Musk's philosophy. As far back as 2014, the Tesla CEO made it clear he had little patience for rigid hierarchies. Information, he argued, should be accessible – regardless of someone's rank. In a company-wide email, Musk encouraged direct communication at all levels. In his words: 'It must be ok for people to talk directly and just make the right thing happen.'

But in spring 2019, not everyone at Tesla seems to agree. One of Lukasz's supervisors has learned about his little email-chat with the CEO – and doesn't like it one bit.

'Who do you think you are?' the manager asks Lukasz. He wants to see his email exchange with Musk. Lukasz refuses. Instead, on April 6, he writes another message to the top.

'After my last email to you, my supervisor/service manager insists on reviewing my correspondence to you. Is that ok with you? Things are getting hot for me. Cannot share anything that could possibly hurt my manager and then possibly me.'

Musk replies the next day – vague, even dismissive. 'I can't read emails unless critical to Tesla. It's literally physically impossible. Your manager cannot be a filter, but I also cannot process emails unless vital to Tesla.'

Lukasz is stunned. What does this mean? Doesn't Musk care if his cars catch fire? If employees are tampering with vehicles in ways that could endanger customers? What happened to doing the right thing? Wasn't that always Musk's rule?

Marked and Marginalized

Lukasz doesn't back down. On April 11, he organizes a meeting with colleagues; a manager joins as well. They go over factory issues – insufficient tools, safety hazards. The next day, the manager assures him by email that all concerns will be 'taken seriously and will be handled in a professional way.'

Lukasz watches and waits. How long will it take for things to actually change? But as time passes, something strange begins to happen: his work laptop starts slowing down badly. Something's off. Lukasz presses CTRL+ALT+DEL, opens Task Manager, and scans the list of running processes. Then he sees the culprit.

One program is eating up more than half the computer's processing power: Code42. A quick Google search sends a chill down his spine. Tesla has installed spyware on his laptop. The maker of Code42 touts it as a tool for identifying employees who pose 'a threat' to their company.

A few days later, Lukasz is transferred – to the basement of the delivery department. He is no longer allowed to talk to his former service team colleagues. When he asks for an explanation, none is given. He's been shut out.

Six days later, Lukasz is called into a meeting. His supervisor is already waiting, along with someone from HR. They tell him there have been complaints. Lukasz is stunned when he receives a written summary of his supposed failures: 'All in all, your behaviour is impacting the work environment in a negative way. [. . .] Using too much time on looking into things if it is correct or not, and [someone] not doing his job.'

On May 29, Lukasz receives a formal warning. He brings in a labor lawyer, and Tesla drops the most serious accusations. A few days later, another warning follows. This time, the company accuses him of 'bad behavior'. As his superior puts it: 'Tesla perceives your claims against colleagues as deliberate misunderstandings in order to waste working time.'

And so, his descent from hero of the day to wrench in the works continues. Lukasz is reassigned again. His colleagues seem to know what's going on – they treat him like a traitor, someone endangering the whole operation. On March 5, 2020, Lukasz sends a final email to Musk. He describes a 'culture of distrust' in Drammen, tells the CEO about management insulting staff and ignoring safety risks.

Musk doesn't reply. Instead, someone from Tesla's compliance department responds: 'We take such concerns seriously and will not tolerate wrongdoing in the workplace.' Lukasz can rest assured, she writes. 'Please be aware that Tesla strictly prohibits any form of retaliation.'

In July, Lukasz receives a negative performance review on Tesla's internal network. 'He has got several complaints about his ability to work well with others, and most will therefore not work with him if asked,' the anonymous assessment reads. Lukasz is labeled uncooperative, not innovative, and untrustworthy in the eyes of his peers.

On September 3, 2020, an HR employee in Oslo writes to a compliance colleague at Tesla's European headquarters in Amsterdam who had looked into Lukasz's claims. In Norway, she writes, there's 'no culture or practice for investigating every single little claim that an employee might have.' She continues: 'We hear stuff all the time that we label as gossip and no one thinks further about it. It doesn't seem right to take a person, who has been given previous warnings for malicious behavior against other employees, so serious, when he reports more employees. I think that would only encourage him to continue.'

According to the staffer, Lukasz isn't part of the solution, but part of the problem. Her conclusion: 'Management is really tired of Lukasz almost to the point that they are angry with him.'

His dream job has become a waking nightmare. Lukasz barely sleeps. He's constantly nauseous. On September 23, 2020, he collapses at work and is rushed to the ER. He's signed off on medical leave. Soon after, he begins therapy.

He could have turned away. Focused on recovery, closed the Tesla chapter. But that's not who Lukasz is. Three years later, we'll speak with his father. 'He was incredibly stubborn and headstrong as a child,' he'll tell us. 'Once Lukasz sets his mind to something, he sees it through – no matter the cost.'

The Reckoning

By the end of 2020, Lukasz has switched to survival mode. He lets go of the idea of improving Tesla and begins looking for ways to fight back. He combs through reports of conflicts others have had with the company. He archives internal emails, chat logs, and Facebook posts. He connects with critics of Elon Musk on social media.

That's how Lukasz comes into contact with Aaron

Greenspan. The American runs a website that publishes court documents on Tesla and other companies. He's widely seen as an enemy of Musk, not least because he bets on Tesla's stock price falling. Musk has sued him.

Lukasz insists he has never shorted Tesla stock, sold information, or profited from the company's flaws. But he spends months exploring them. Greenspan's online archive proves to be a valuable resource.

On the evening of November 2, 2021, Lukasz is sitting at his desk, work laptop open in front of him. He logs into Tesla's IT system, clicking here and there. Then he stumbles across a file that appears to originate from the company's finance department. He tries to open it. It works. Lukasz can hardly believe it. He closes the file and opens another – same result. As a service technician, he shouldn't have access to any of this. But he does.

Gripped by a sort of search frenzy, Lukasz doesn't notice that midnight has come and gone (and it is now his 36th birthday). He scours invoices and uncovers pricing structures from Tesla's business partners. The company's financial infrastructure lies open before him.

At 6:27 a.m., he finally stops. Before closing the laptop, he sends a short video message to Greenspan. Lukasz is unshaven, the screen reflected in his glasses, dark circles under his eyes. Staring into the camera, he says: 'Fuck. I'm a zombie.'

A zombie on a collision course with Elon Musk.

Now Lukasz begins collecting all the evidence he can find of wrongdoing at Tesla. Aware that the spy software Code42 is still running on his machine, he initially avoids taking screenshots or downloading files. Instead, he photographs his computer screen – hundreds of images soon fill up his iPhone. But the fear gradually gives way to convenience. Lukasz starts taking screenshots. Eventually, he even begins downloading files. At any moment, he expects Tesla to revoke his access or send someone knocking on his door.

Nothing happens.

On November 12, 2021, Lukasz and Greenspan submit a 21-page complaint to the U.S. Securities and Exchange Commission. They accuse Tesla of poor data security and a lack of transparency toward investors. Lukasz hopes to have gained a powerful ally in his battle with Musk. He believes it's only a matter of time before the SEC holds the company accountable.

But the American authorities stay silent. Instead of hearing from the SEC, Lukasz gets a letter from Tesla. The company has reviewed his case, a human resources staffer writes on December 14, 2021. She has bad Christmas news.

'Hi Lukasz, as you can see from the attachment, we have unfortunately decided to terminate your position as an Associate Service Technician. The notice of dismissal will be sent by registered letter. The decision around dismissal of your employment has been reached after a thorough consideration and assessment of your case. Again, we are sorry for the inconvenience this may have for you. We wish you all the best!'

Lukasz sees it as confirmation. Tesla doesn't want to hear his warnings. It doesn't want to hear him at all. Two and a half years have passed since he reached into the burning Model 3 in Lillestrøm. His efforts to help improve the company have only brought him trouble. He now sees Tesla as incapable of correcting its course.

After the SEC, Lukasz turns to the U.S. National Highway Traffic Safety Administration and shares even more explosive information. On January 23, 2022, he uncovers a massive security risk when he opens an Excel file titled 'Headcount Master'. It contains detailed data on Tesla employees worldwide, including salaries for nearly 100,000 staff. Lukasz is stunned that he can access the file, apparently without raising any red flags in the system.

And so, he continues. He finds documents listing termination reasons, private phone numbers, social security numbers, architectural blueprints, and legal records. Even queries about Elon Musk's personal expenses – and those of his bodyguards – return results without restriction.

Lukasz's data archive now exceeds 100 gigabytes. He informs additional authorities in the U.S. and Europe and reaches out to journalists. Every week, he braces for the big bang.

It never comes.

A handful of media outlets touch on some of his findings. Some regulators launch inquiries, but clear consequences remain absent. For Lukasz, 2022 becomes a year of quiet frustration – a relentless test of his patience.

In November, he finally reaches out to *Handelsblatt*. To us.

He's cautious. Six weeks and countless encrypted messages pass before he reveals his name. After a tense demonstration of the data breach in Drammen, Norway, we begin organizing the material Lukasz has compiled. At *Handelsblatt*, Team Tesla is born.

Starting in February 2023, 13 reporters reach out to more than 200 Tesla customers and employees. Some of them are so alarmed that they contact the headquarters in the U.S., wanting to know how German journalists got hold of their personal data.

The fuse is burning. Will it reach the powder keg?

Operation Handelsblatt

Tesla moves into crisis mode. On March 9, 2023, the company activates its 'Security Intelligence Investigations' team, a unit whose daily routine involves monitoring employees and plugging leaks as quickly as possible. The new mission:

track down the whistleblower feeding inside information to *Handelsblatt*.

The five-member team includes former intelligence operatives, seasoned investigators, and digital forensics specialists – people Tesla pays up to $200,000 a year. Among them is a former CIA officer who, on her LinkedIn profile, touts her experience with 'the highest tier of government' and her expertise in 'insider threats.' Once tasked with tracking spies and terrorists, she is now tracking down a whistleblower in Norway. She's hunting Lukasz.

Internally, the case is listed under file number 23-0081, categorized as 'Media Leaks,' and labeled '*Handelsblatt* Outreach.' But even this high-level unit can no longer seal the leak. Since November 2022, we've analyzed more than 23,000 Tesla documents and, on May 10, 2023, sent the company a list of 65 critical questions.

On May 15, 2023, Elon Musk sends a terse directive to his staff. 'I would like to gain a better understanding of our hiring,' he writes. From now on, he wants to receive a weekly list of all incoming Tesla employees. 'Think carefully before sending me a request,' Musk adds. 'No one can join Tesla, even as a contractor, until you receive my email approval.'

The next day, Tesla contacts us. Attorney Joseph Alm threatens legal action against the whistleblower and demands we delete the documents Lukasz provided. At the same time, he refuses to answer any of our questions. On May 26, *Handelsblatt* publishes a ten-page exposé: 'The Tesla Files'.

Our reporting catches the attention of regulators. If the allegations prove true, it would be particularly serious from a data privacy perspective, says Dagmar Hartge, data protection commissioner for Brandenburg – the German state where Tesla's factory is located. The lead authority for the European investigation is the Dutch data protection agency, Autoriteit Persoonsgegevens, based in Amsterdam, where Tesla's European headquarters is registered. The Dutch authority relays the material from 'The Tesla Files' to its counterparts via the IMI-Net internal reporting system. Tesla's problem now begins to cross borders – one country after another.

Then Tesla tries to flip the script. One weekend in late May, we receive more than 100 emails from Tesla employees demanding that we delete their personal data. All use the exact same language. We later learn that Tesla had provided staff with a template and encouraged them to contact us. By Monday morning, there are already 150 emails. More arrive by the minute. By noon, it's 500. By evening, more than 1,000.

We reply to each message, thanking the sender and explaining why we cannot comply: under data protection law, there is no right to deletion of information used in legitimate journalistic investigations. Otherwise, politicians and corporate leaders could prevent coverage about themselves simply by demanding that all data be erased. A few employees write back a second time. Then this front falls silent.

As we respond to emails from Tesla staff, Musk's secret agents dissect our work. Line by line, they cross-reference each fact in our articles with the company's internal files. Using Tesla's log data, they narrow down the list of employees who accessed these documents at specific times.

Eventually, they identify a single account: 'lkrupski.' On June 1, 2023, Lukasz feels the full consequences of going public. Still riding the wave of media attention, he's caught off-guard when Norwegian police knock on his door, pull him out of bed, and search his apartment.

His data is now considered evidence, and an independent expert will examine the files. Then, a judge will decide what Tesla can and cannot use against him.

Has Netflix Called Yet?

While Lukasz awaits his trial, his revelations continue to pile pressure on Tesla. On July 21, 2023, the company explicitly references our reporting in its quarterly statement. Tesla lists 'The Tesla Files' alongside investigations by the U.S. Securities and Exchange Commission and the U.S. Department of Justice.

'A foreign news outlet reported that it obtained certain misappropriated data including, purportedly, among other things, non-public Tesla business and personal information. While Tesla's investigation remains ongoing, we are working with certain law enforcement and other authorities,' the company states. The warning to investors is clear: 'An unfavorable ruling or development [. . .] could materially harm our business, financial standing, or brand.'

Former employees now seek accountability for Tesla's alleged data protection failures. At least two class-action lawsuits are pending in courts in California and Pennsylvania. One plaintiff, Benson Pai, formerly employed at Tesla's Fremont plant, demands $100 million in damages. In his complaint, Pai directly cites the allegations documented by *Handelsblatt*, accusing Tesla of failing to take 'standard and reasonably available steps' to prevent the data breach, failing to 'monitor and timely detect the Data Breach' and failing to provide 'prompt and accurate notice.'

Indeed, while Tesla informed European authorities of the leak back in May, it delayed for months before notifying its own staff. On August 18, 2023, more than 75,000 employees receive a message. Tesla wants to assure them it is taking appropriate measures concerning 'The Tesla Files' – including offering affected employees membership in a service that assists potential victims of identity theft. The primary concern: the social security numbers

Lukasz found in an Excel spreadsheet. To criminals, these numbers represent ideal tools to cause enormous damage.

A month later, Tesla finds a scapegoat. Jason Smith, an employee of more than seven years, served as the lead administrator for Jira, the project management system from which Lukasz extracted the Tesla Files. For Smith, who is preparing for paternity leave, the termination comes out of nowhere. 'Last Friday I was, to my complete surprise and shock, fired from Tesla,' he writes on LinkedIn. 'I was blamed for a situation that was not in my set of responsibilities when an incident happened.'

Smith describes the termination as a deep personal wound. He loved his job and is shaken by Tesla's approach. More than 800 users react to his post; nearly 70 comment, offering comfort and sharing similar experiences.

When asked what had happened, Smith summed it up simply: 'The long story short is that I was held "accountable" because a user leaked information from Jira,' Smith replies. He calls it ridiculous to blame him for the data leak. 'The sad/frustrating part is I raised concerns about the risk of that several years ago and those concerns were ignored.'

How will regulators respond to Tesla's handling of customer, employee, and business-partner data? In the United States, the responsibility for enforcing data protection laws lies with the state attorneys general. Each of the 50 states can launch its own investigation if its residents are affected. Half a dozen authorities are currently examining the case.

In Connecticut, the attorney general's office states it will closely monitor Tesla's handling of the leak. Indiana regulators reach out to *Handelsblatt*, asking for further details about the data breach to help their citizens 'in the best possible way.' Aaron D. Ford, the attorney general of Nevada, emphasizes in a statement that protecting identities is not only a personal duty but also a corporate obligation.

At *Handelsblatt*, we continue publishing. Since our first story in May, our access to Tesla has completely changed. New whistleblowers have come forward. At the German Gigafactory in Grünheide, previously almost hermetically sealed to journalists, we've established a row of reliable contacts. And we continue mining the Tesla Files – 100 gigabytes can be an almost inexhaustible resource.

Within our newsroom, we're known as the Tesla Crew. Everyone loves this story – it simply has it all. A flashy corporation, seemingly dangerous products, the world's richest man, secret agents hunting our informant.

Colleagues keep asking me: 'Has Netflix called yet?'

At first, I laugh. Then I start to wonder. Ever since streaming services reshaped the media landscape, documentaries – even those on business topics – have become hot commodities. Series like Dopesick, about Purdue Pharma, and The Dropout, chronicling fraudster Elizabeth Holmes and her company Theranos, have captivated millions. Couldn't the Tesla Files also make for gripping viewing?

On a whim, I reach out to Netflix, gauging their interest. Emails lead to video calls, which swiftly lead to a connection with a film studio. My colleagues were clearly onto something: the film producer is instantly captivated by 'The Tesla Files'. Soon we arrange an introductory meeting with Lukasz. A plan begins to form: the whistleblower no longer wants to remain hidden. When the time is right, Lukasz will take his story public.

In early September 2023, a film crew flies to Oslo for initial shooting. Back in Düsseldorf, we begin extensive research for a comprehensive profile of Lukasz. Over the following weeks, we bombard him with endless questions. At the same time, his lawyer suggests bringing an American newspaper onboard. If Lukasz is going to show his face, why not on Musk's own turf? I call a journalist at the *New York Times*, a colleague I've known for years. It doesn't take much convincing.

Two weeks later, I'm back in Oslo. Lukasz and I speak for hours about what has transpired in recent weeks – and what still lies ahead. He seems uncertain. The former Tesla colleague who helped him secure data has broken off contact and no longer responds to messages. Lukasz's loneliness has deepened. After two days, I have to return home.

At the office, we begin to reconstruct Lukasz's life – year by year, month by month. I keep a near-constant hotline to Norway. Hundreds of small details need clarification. Thankfully, Lukasz is meticulous. His archive of emails, chat messages, and photos is scattered across multiple devices, yet he usually locates documentation quickly, supporting nearly every detail.

Not everything in our investigation goes smoothly. At first, Lukasz willingly shares details about his childhood, speaking openly about his two sisters and his brother, each born two years apart. But there's a darker side to his family history, one that required therapy. This remains hidden from us.

On September 25, I ask Lukasz if I might speak to his parents. He is skeptical and hesitates, but eventually sends an email address. A few days later, we have their phone number. I pass it on to my colleague Roman Tyborski, who speaks fluent Polish.

Lukasz's father, Teodor Krupski, describes his son as serious, introverted, and precociously talented – with an aura of mystery even as a child. Father and son rarely see each other nowadays. 'We write occasionally,' says Krupski Sr., 'but I don't interfere in Lukasz's life. He's an adult and makes his own choices.'

Slowly, the text takes shape. Details accumulate. Only now do I realize Lukasz's compulsive thoroughness whenever he's determined to act. At Tesla, he documented everything suspicious. Faulty equipment, spilled liquids in the workspace, screenshots of chat messages where colleagues made racist or sexist remarks. Even their Facebook profiles made it to his archive.

Gradually, fewer gaps remain in Lukasz's story, but some points still require clarification. I press further until Lukasz suddenly explodes. 'I made that up,' he snaps. 'I also sold the information to shortsellers, Chinese, Saudi and Russian governments.'

I get what's happening. Lukasz is tense, overburdened. The endless barrage of my questions feels like mistrust. I carefully explain that it's my job to scrutinize details. He replies curtly that he no longer cares about articles focusing on him. I apologize and give him space over the weekend.

By Tuesday, we resume contact as though nothing has happened. We discuss U.S. attorneys potentially interested in his case. I send him the interview transcript we worked on previously. His tone darkens again. 'I wonder if it matters,' he writes on October 10, 2023. 'I am so exhausted. And since they have more resources they can shout and no one will hear me. They will arrest me, shoot me, or place a bomb under my car.'

The next day, he sends me corrections to our interview – and offers to scan old family photos for the article. 'If there is something I can do,' he says, 'I will try to do.'

Years in this job have taught me how volatile whistle-blowers can be. Usually under immense stress, they've challenged opponents a thousand times more powerful. Their frustration inevitably spills onto those who constantly probe for more details. Relationships between journalists and sources are fragile. Whistleblowers often expect allies. But we're not their allies – we're their chroniclers. And when those expectations break, disappointment sets in. All I have is patience and calm.

Lukasz Fights On

Tesla has demanded Lukasz delete all data and, in return, offers to drop every accusation against him. His lawyer considers this a reasonable course of action, but Lukasz refuses to settle – not now, not on the eve of going public as a whistleblower. He questions whether his attorney truly represents his interests or, rather, those of the law firm. 'I am seeing my doctor later. Because I honestly don't know if I am going nuts,' he writes me in a message, 'or I am being f****d by my own people.'

The conflict with his lawyer drags on for several days – until Lukasz comes to Düsseldorf to visit *Handelsblatt*. It is Thursday, November 2, 2023. Over the past few days, we have worked virtually nonstop on our piece, weaving Lukasz's story tightly into the broader Tesla narrative. Tomorrow, his face will appear on the front page of *Handelsblatt*. We take photographs, give our guest a tour of the newsroom, and record a podcast.

For me, these final hours bring relief. Lukasz, meanwhile, grows increasingly anxious. Michael and our colleagues meticulously verify every figure, every caption. Finally, around six in the evening, we give Lukasz the thumbs-up. The paper is being printed. We go out together to a nearby restaurant. Later, we return to the publishing house. The first copies are being delivered. As Lukasz holds a freshly printed newspaper, all the doubts and fears of recent days fall away. Later, as I drop him off at his hotel, we embrace goodbye. He is happy.

The next day feels like the cool-down after a race. Months of research have culminated in one monumental story. Lukasz returns to the newsroom; it's his birthday. We have flowers and cake. Our editor-in-chief, Sebastian Matthes, congratulates Lukasz on his 38th birthday and chats warmly with him. Following our enormous efforts, fatigue sets in. It's Friday evening; my colleagues long to rejoin their families, sorely neglected of late. At the end of the day, Lukasz and I take the tram back to my home.

My wife Ayuk and our two-year-old daughter Melinda are waiting there. Dylan, my eldest son, is still at football practice, while his brother Brandon sits with his friend Jonas at the Xbox, steering a passenger plane across the Atlantic. Brandon has disappeared into the skies lately – the flight sim is all he talks about.

Ayuk greets us warmly, then heads out shortly afterward for an appointment of her own. After his friend leaves,

Brandon eagerly talks to Lukasz about the advantages and drawbacks of various aircraft. The two click.

We order Greek takeout from the place next door. Dylan returns from training, hungry as a bear. I watch Lukasz, sitting comfortably among my children. In our countless conversations over recent months, he's confided to me how much he once thought about having his own family. But for four long years, his struggle with Tesla has blocked out everything else.

His birthday is nearly over. I walk Lukasz to the tram, which will take him directly to his hotel. We won't see each other again before his flight the next day. His plane will be taking off just as I'm photographing Brandon's football match. He'll probably already be on his sofa back in Drammen when I take Dylan to his game. As we say our goodbyes, I thank him, and he thanks me.

Now, Lukasz wages his legal fight as a public whistle-blower. At the same time as *Handelsblatt*, the Norwegian paper *Dagens Næringsliv* publishes a major profile of him. One week later, the *New York Times* follows suit with the headline: 'Man vs. Musk.' On December 4, the international nonprofit Blueprint for Free Speech awards Lukasz for his contribution to freedom of expression. Another wave of media coverage ensues. Lukasz begins preparing his own lawsuit against Tesla.

In the spring of 2024, mediation takes place. Despite initial harsh threats, Tesla is now ready to waive all claims against Lukasz. The U.S. company even offers to compensate him for the lost earnings during the years spent battling Tesla. Their only condition: Lukasz must delete the Tesla Files. He rejects the offer. He wants a court decision.

Then, in December 2024, Lukasz finally wins the vindication he's fought for all along. The Norwegian District Court of Buskerud rules that Tesla treated Krupski unlawfully and violated his rights as a whistleblower. Tesla must pay its former service technician €10,000 in compensation, plus €170,000 in legal costs.

News of the verdict quickly spreads. A fellow journalist asks us for a comment for his own coverage. He wants to know what the judgment means for *Handelsblatt*: 'Are you also relieved by this?'

We decide not to give a statement. However deeply involved we have become in this case, we're not party to it. Yet we naturally reach out to Lukasz, asking how he now feels. Rehabilitated? Freed? Happy? 'I've become calmer,' he replies. 'I'm at peace with myself. I sleep more and try to sort everything out. Now I can finally focus on rebuilding my life.'

But Tesla has other plans. The new year arrives, but as of mid-January, Lukasz is still waiting in vain for the promised compensation. On January 22, 2025, Tesla files an appeal. His former employer now demands release from all claims and goes even further: Tesla's lawyers insist 'Lukasz Teodor Krupski is liable for the financial loss suffered by Tesla Norway AS and/or other Tesla Group companies as a result of his unauthorized acquisition and sharing of information belonging to and/or taken from Tesla Norway AS and/or other companies in the Tesla Group.' Additionally, Lukasz must bear all procedural costs. 'Omissions, distortions. The usual,' Lukasz comments on the letter. 'It's like dealing with a madman.'

A friend forwards him a message from Tesla's internal communication system. The subject line reads 'Jira at Tesla' – the very project management system whose careless handling Lukasz identified as a massive security risk. Nearly all the documents now known as the Tesla Files had lain unprotected in Jira.

Tesla's message reads like a confession. At the request of the legal department, access rights in Jira will now be strictly controlled, the carmaker informs its staff. Furthermore, it is now forbidden to store sensitive information like customer addresses, salary details, or social security numbers in the system.

Lukasz can't help but smile as he reads these words. His warnings, once ignored, are now etched into company policy. With his revelations printed across newspapers, Tesla tells its employees: 'If you see something, say something.'

CHAPTER 7

Elon's DNA

The first thing new Tesla employees learn isn't how to build cars. It's how to think like Elon Musk. In 2020, the company unveiled a training presentation titled 'The DNA of Elon and Applying it to Daily Work.' At its center: a surreal collage. The outer ring is filled with buzzwords – transformation, intelligence, ambition. The middle: logos of Musk's companies. And in the core: a photo of Musk himself, chin lifted, eyes set on the future. A man as mission statement.

When we opened the Tesla Files, our first instinct was simple: search for Musk. His name. His aliases. The people around him. The trail quickly led to surprising places. We found references to his ex-girlfriends, internal nick-names, and even his social security number buried in a spreadsheet.

The deeper we dug, the clearer the picture became: Tesla isn't run like a traditional company. It's run like a command center – with Musk at the core.

His bodyguards call him Voyager, a nod to the NASA probe. Employees refer to him as N1 – Number One. They describe a man who micromanages obsessively, rewrites rules at will, and sees betrayal lurking around every corner. Former staffers tell us what really scares people inside Tesla: not failure, not layoffs. His anger.

Musk doesn't just lead. He overrides. Managers, systems, rules – if they slow him down, they're bypassed. Even when they were his idea to begin with.

To understand Elon Musk's vision of leadership, one must look at what he expects from others: total replication. The DNA presentation lays it out in stark terms. The ideal Tesla employee? A clone of the man at the top.

At this point, it's hard not to think of Musk's children – all 14 of them. He speaks about reproduction not just as a personal choice but as a mission – even a duty. 'If birth rates continue to plummet, human civilization will end,' Musk wrote on April 29, 2024, on X. Seven months later, he added: 'Instead of teaching fear of pregnancy, we should teach fear of childlessness.' Musk has said he 'wants smart people to have kids.'

Apparently, he can't find anyone smarter than himself. According to an article in the *Wall Street Journal*, Musk refers to his offspring as a 'legion.' Historically, the term described one of the most powerful, professional, and feared military units of the ancient world – a key factor in the Roman Empire's vast expansion and dominance.

Just as the Roman legion played a pivotal role in shaping history, Musk appears to view his growing number of children as part of a larger mission to shape the future. In February 2025, it became public that Musk had fathered his 14th child, a son named Romulus. His mother is conservative influencer Ashley St. Clair. The two met over X and later in person while vacationing on the Caribbean island of Saint Barthélemy in 2024.

Romulus seems to be a statement as much as a son. Named after the mythical son of the god Mars, who became the legendary founder of Rome, he reflects Musk's growing fascination with the ancient empire – renowned for its endless hunger for power and reach.

During St. Clair's pregnancy, Musk reportedly suggested bringing in other women to accelerate the process of having more children. 'To reach legion-level before the apocalypse,' he wrote to St. Clair in a text message viewed by *The Wall Street Journal*, 'we will need to use surrogates.'

The ideological stance behind this is called pronatalism. The worldview promotes reproduction and encourages people to have more children, often viewing childbirth as a social, moral, or even existential imperative. At its core, pronatalism is based on the belief that a growing population is beneficial – whether for economic growth, societal survival, or ensuring the long-term future of humanity.

In its more extreme forms, pronatalism intersects with ideas about legacy, genetics, and influence – where having many children is seen as a way to shape the future or ensure the continuity of one's values, identity, or, in Musk's case, even DNA. The Tesla CEO seems to view both his companies and his children as integral parts of his legacy, each helping to push his visions forward. The Tesla presentation outlining what defines Musk reads like a kind of corporate pronatalism.

According to the slides, Musk is defined by a 'mixture of high intelligence and ambition.' He embodies 'first principles thinking,' approaches problems 'from a physics framework,' and solves them using 'basic math.' He has a 'quick understanding of things on a very detailed level,' wastes no time on tasks without value, and tackles the hardest problems first – often before others are even awake.

Routine work is handled 'on the fly.' Meetings? No more than 15 minutes. The document paints a picture of a leader as stripped-down and optimized as the robots he hopes one day to build.

Tesla expects its leaders to get the absolute most out of their teams – and themselves. That's why managers are trained in a set of so-called 'manager fundamentals,' presented in internal workshops. The premise: you can't lead others until you learn to lead yourself. Participants are asked to reflect on their own personalities – identify strengths, acknowledge weaknesses, and align both toward one goal: Tesla's growth.

One workshop module stands out. It's called 'The Accidental Jerk'. 'Even with the best of intentions, we all sometimes diminish others without realising,' the materials state. 'This is your inner jerk.' The module guides managers through exercises designed to recognize when that tendency shows up – 'put your inner jerk into place' and adopt 'more useful behaviour that will improve your effectiveness as a leader.' It's part therapy, part self-optimization – empathy engineered for efficiency.

Other modules sketch out leadership types to avoid. 'The Tyrant,' who creates a stressful atmosphere that blocks thinking. 'The Micromanager,' who 'drives results through personal involvement.' 'The Know-It-All,' constantly issuing directives that show off how much they know. 'The Empire Builder,' who 'hordes resources and underutilises talent.'

All of them, the materials say, share one toxic belief: 'People won't figure it out without me.' They are the narcissists 'who want to be the smartest person in the room, diminish others, [and] control their team members.' According to current and former employees, all of these traits apply to Musk himself.

And yet – ironically – the materials also quote him telling his managers: 'Don't be a jerk.'

How do Tesla's managers know which of their traits need work? A training module titled 'Who Am I as a Leader?' pushes them to reflect on that very question. It asks them to examine their personality: What are your strengths? Where might those strengths become liabilities? And – most importantly – how do you stop that from happening?

'You can't lead others unless you have a strong sense of who you are and what you stand for,' the presentation says. 'But understanding yourself can be difficult, when there might be things that we don't even know about ourselves.'

To help managers bridge that gap, Tesla turns to a psychological model known as the Johari Window. It maps out self-perception in four fields: traits known to both you and others; private traits only you know; unknown traits that no one sees; and blind spots – the parts of yourself that can only be revealed through outside feedback.

Tesla encourages its leaders to stay open, vulnerable, coachable. That's the only way to shrink those blind spots. But there's a catch: you can't get honest feedback without trust. 'Only when there's trust, people feel like they can share and give you honest feedback,' the materials explain. Then they pose the question every manager is meant to wrestle with: 'How can you build trust within the team?'

Trust, in Tesla's framework, is a prerequisite for feedback and growth. Managers are told to cultivate it – to be open, receptive, and self-aware. But what happens when that principle is tested at the top? What does trust look like when the person in question is Elon Musk himself? One barely noticed moment captures the answer with striking clarity.

In October 2023, during an earnings call, Musk talks about Tesla's bright future. But there's one problem: no one can hear him. His microphone is muted.

He keeps talking. And talking. Unaware that his words are going nowhere. On their screens, analysts can do little more than read his lips. Time passes. Still, no one interrupts. No one trusts that speaking up would be a good idea.

Eventually, someone at Tesla unmutes him. But no one dares to point out the issue. Musk continues as if nothing happened – he doesn't repeat the start of his statement. In the transcript, where his opening should be, it now reads: '[Audio gap]' or '[Call starts abruptly].'

A statement from the richest man on Earth turns into a farce – because no one speaks the obvious truth: he can't be heard. The moment reveals just how deeply fear shapes the environment around him. Even during a routine earnings call, silence wins.

The episode recalls *The Emperor's New Clothes*, a fairy tale by Danish author Hans Christian Andersen. In his story, no one dares to tell the ruler what's plainly obvious: that he's not wearing any clothes. Everyone stays silent – until a child speaks the truth.

There were no children at that Tesla conference call. And Musk, it seems, surrounds himself with yes-men. The open feedback culture he promotes? If it exists at Tesla, it stops at the door to his office.

The Paranoid Billionaire

The distrust between Musk and his employees runs both ways. Inside Tesla, he is known as a control freak – prone to suspicion, sudden turns, and flashes of paranoia.

In 2018, he startles staff with an internal email. An employee, he writes, has committed 'quite extensive and damaging sabotage.' Musk adds: 'The full extent of his actions are not yet clear, but what he has admitted to so far is pretty bad.' Then he offers a warning: 'There may be considerably more to this situation than meets the eye.'

Musk asks his team to consider who the saboteur might have been working with. 'As you know, there are a long list of organizations that want Tesla to die.' Among them: shortsellers on Wall Street, 'who have already lost billions of dollars and stand to lose a lot more.' Also on the list: oil and gas companies – 'They don't love the idea of Tesla advancing the progress of solar power & electric cars. Don't want to blow your mind, but rumor has it that those companies are sometimes not super nice.'

Then Musk points to a third group: traditional car manufacturers. 'If they're willing to cheat so much about emissions, maybe they're willing to cheat in other ways?' He puts his workers on high alert, asking them to 'please be extremely vigilant, particularly over the next few weeks as we ramp up the production rate to 5k/week. This is when outside forces have the strongest motivation to stop us.'

As the Japanese say: business is war. And in war, there are casualties. By May 2022, Musk seems to believe anything is possible. He tweets: 'If I die under mysterious circumstances, it's been nice knowin ya.'

In early March 2024, unknown individuals carry out an arson attack on a power line near Grünheide, cutting electricity to Tesla's German Gigafactory. Not long after, Musk visits the site. Footage shared by employees shows him on stage, visibly agitated.

'I think somebody should be arrested because it's like, they just did like major arson,' he says. 'That was really like not cool at all. Not cool.'

Musk calls on authorities to 'improve security,' and announces he plans to meet with officials to push for action. 'We have to make sure that somebody, you know, they find the criminals that did this and that they must have some kind of punishment.'

Then, almost offhandedly, he adds a line that hangs in the air: 'I mean, sometimes when I see something that seems like so ridiculous, I sort of wonder, are there larger forces at work? Like, are they just puppets and someone else is behind it?'

He ends by saying he's off to meet with the authorities – to, as he puts it, 'track down the terrorists.'

In June 2024, Musk addresses Tesla shareholders. 'I mean, it is getting a little crazy these days,' he says. 'The probability that a homicidal maniac will try to kill you is proportionate to how many homicidal maniacs hear your name. They hear my name a lot, so I'm like, "OK, I'm on the list," you know.'

A month later, just after the assassination attempt on Donald Trump, Musk posts on X: 'Dangerous times ahead. Two people (separate occasions) have already tried to kill me in the past eight months. They were arrested with guns about 20 mins drive from Tesla HQ in Texas.'

Musk is prepared. Invoices from the Tesla Files show that his bodyguards accompany him everywhere – 24 hours a day. They handle his travel, pay his hotel bills, shop for him, bring him medicine. One invoice cites protection from 'inappropriate pursuers.'

Another lists $312 for 'souvenirs for Voyager's kids.' A different one totals over $20,000 – for a 'confidential investigation' into the former owner of the domain www.tesla.com.

In January 2016, Musk travels to Mexico, Hong Kong, London, Paris, Israel, and Texas. The associated bill totals over $163,000. The cost is split between Musk, Tesla, and SpaceX. Not long after, Musk founds Foundation Security – referencing his favorite sci-fi series, *Foundation* by Isaac Asimov. According to documents in the Tesla Files, the company is run by Gavin de Becker. Its stated purpose: 'private business with internal security staff.'

In December 2023, Tesla announces a formal contract with a security company owned by Musk – one the automaker has already paid several million dollars. From then on, Musk profits twice from his own protection.

Inside Tesla, employees describe a company with two castes: Musk, and everyone else.

Tesla fires workers for using cannabis. Musk smokes a joint on camera – and when asked if that's OK, he doesn't blink. While Tesla prohibits weapons on its campuses, Musk campaigns on X for the 'right to bear arms' to prevent the government from 'taking people's rights away.'

Visitors to Tesla must sign nondisclosure agreements. Employees whose guests take photos of the factory face 'disciplinary action.' Musk's mother, Maye, posts photos of her 'incredible visit' to the Grünheide plant.

Tesla touts its diversity and stresses in official policies that it does not tolerate discrimination. At the same time, Musk is campaigning for the German right-wing party Alternative für Deutschland (AfD) and its leader, Alice Weidel, who believes that 'political correctness belongs in the trash heap of history.'

Cars and Politics

'Only the AfD can save Germany,' Musk writes on his platform X on December 20, 2024. Germany is in a government crisis. The AfD is rising. Critics accuse Musk of backing an extremist party. Inside Tesla's Grünheide factory, a different question surfaces: Does the boss even know who he is endorsing? How does his support align with the super-rational, numbers-driven leader from the DNA presentation – especially given the party's history with Tesla?

If the AfD had its way, the Grünheide factory would have never even existed. The approval process for the factory began in early 2020. By February, construction was underway. 'It's the wrong choice of location,' said Kathleen Muxel, an AfD member in Brandenburg's state parliament, at a rally in June. The AfD called for a halt to the build.

Two months later, the party protested in Grünheide, condemning Musk's plans as 'a nightmare for people and nature.' Other AfD voices warned of groundwater damage and the 'lack of long-term viability of electric vehicles.' When construction advanced, Muxel bemoaned it as 'a disgrace to the German rule of law.'

To the AfD, fighting against Tesla was fighting for Germany. 'Who does Tesla think they are? Where would we end up if we let every "Mister Billionaire" from America destroy our homeland at will?' said Muxel in April 2021. Musk, of course, pushed through. By March 2022, the first cars were rolling out of the shiny new Gigafactory – the crown jewel of Tesla, as he called it.

By June 2023, 12,000 people were working there. The AfD remained hostile. What others hailed as a success, the party called 'assistance in the destruction of the domestic German automobile industry by rolling out a "red carpet" of subsidies for an American company.' The international makeup of the workforce didn't sit well either. On its website, the AfD wrote: 'They are only partly from Brandenburg, mostly from Berlin, and don't seem to be locals based on their appearance.'

Musk showed little reaction. After sharing a video that criticized German lifeboat crews and expressed hope 'that the AfD wins the elections to stop this European suicide,' his response to the backlash was: 'I have not "supported" any political party and don't know AfD from a hole in the ground.'

It's unclear when – or if – that changed. In September 2024, just before state elections, factory director André Thierig addressed employees. 'There are parties that are obviously against us,' he said, reminding his staff of the AfD mantra: Electric cars are a bad idea and Germany was better off without Tesla. Three days later, the AfD became the second-largest party in the state of Brandenburg, with 29.2 percent of the vote.

Musk seemed unaware. As the national elections in Germany loomed, he wrote in an op-ed: 'As someone who has made significant investments in the German industrial and technological landscape, I believe I have the right to speak openly about my political orientation.' Germany, Musk argued, needed a party that allowed companies like Tesla to thrive. 'The AfD is the only party that opens this path.'

Surreal as all this may have been, the AfD got the message. After years of trying to shut Tesla down, its representatives changed their tune. As Steffen John, the party's economic and labor policy spokesperson in the region said: 'Tesla needs planning and supply security.'

Still, the dilemma remained. Tesla had always prided itself on the many nations that work side by side in the Gigafactory. Now, Musk was supporting a party hostile to the multi-national workforce employed at Tesla and classified by the federal authorities as 'far-right extremist'.

We inquire how this aligns with Tesla's values. We ask management, but receive no response. We find out that some employees tried to put the topic up for discussion at the worker's council, but were voted down. Even on the internal chat boards, there is no chatter about the AfD. Insiders tell us the fear is too great that even raising the issue might backfire. No one even wants to create the impression of criticizing Elon Musk.

'How is anyone supposed to make sense of that?' one staffer asks when we bring it up during our reporting. Another employee, reflecting on colleagues who still admire Musk, says: 'Sometimes you get the feeling they don't even realize what he's doing.'

A Godlike Boss

How does Musk imagine the communication process with his employees? In an email dated October 4, 2021, he lays out three options for how managers should respond to his instructions if they have any doubts. First: explain why it's wrong. Second: ask for clarification if the instruction seems ambiguous. Or third: obey and execute. Musk writes: 'If none of the above are done, that manager will be asked to resign immediately.'

It's also revealing to look at Musk's direct reports – the people who answer to him personally. They are, in the vast majority, male, white, and engineers. Of more than 35 individuals who have reported directly to Musk at Tesla in recent years, only five are women. More than half have technical backgrounds.

In the summer of 2022, the Tesla Files list 22 direct reports. But the structure has changed. Their power is diluted, their influence limited. There are more of them now – but still only one person at the top. Even as Musk runs half a dozen companies, meddles in geopolitics, and spends half the night glued to video games – everything still flows through him.

Tesla is Elon. Elon is Tesla.

That's the mantra we keep hearing when we speak to employees. Some say it with a trace of admiration. Others, with something closer to exhaustion.

To some, Musk is a role model. To others, a figure of blind idealization.

'I always compare it to a cult. Everyone orients themselves around the boss, like a divine leader figure,' says an employee from the plant in Grünheide, Germany. Another describes it as 'a mix between Henry Ford and Scientology.'

Internally, employees speak of 'arbitrariness from the top' and Musk's 'bullheadedness.' His decisions, they say, often humiliate the workforce. Tesla doesn't want thinkers – it wants 'henchmen and trained monkeys,' says a former service center manager. 'There's zero room for independent decisions.'

Employees describe what happens when Musk calls a meeting – out of nowhere – at 1 a.m. on a weekend. They scramble to log on; half-asleep, in front of laptops or under fluorescent lights at the office, they brace themselves. Sometimes, they're called idiots.

On another occasion, Musk walks through the Palo Alto offices and takes issue with the sight of trash cans.

Every employee is ordered to remove theirs. From then on, all trash has to be disposed of in shared containers in the copy room – out of Musk's sight.

It is a snapshot of the Tesla way. When Musk has an idea, it's implemented immediately.

And even those who might know better choose to say nothing.

Frequent side effects of absolute power include favoritism and nepotism. At Tesla, proximity to Musk is currency. Just as his mother receives special treatment during her visits, his acquaintances, too, can expect preferential service.

In January 2019, when Musk's neighbor is told she'll have to wait two months for repairs on her Model S, a Tesla service advisor in California writes: 'Will be doing everything to get the vehicle completed sooner.' A few months later, the same advisor opens a ticket to fast-track repairs for a New York attorney. At the top of the entry: 'Customer states he is friends with Elon Musk.'

The Tesla Files are filled with signs that anything tied to Musk is assigned top priority. Sometimes it's a direct order – like in December 2020, when as many Model Ys as possible are to be delivered at short notice.

Sometimes, it's a whim: Musk wants to add a secret Easter egg; if a driver activates Autopilot four times in quick succession, the screen should display the final level of Mario Kart – complete with 'Don't Fear the Reaper' by the American rock band Blue Öyster Cult.

Internally, employees constantly brace for what they call an 'Elon escalation' – a kind of code red. To avoid triggering one, managers sometimes pull strings or call in favors. When they don't have that option, desperation sets in. In a ticket related to an internal IT system, an employee ends the to-do list with a final line: 'More prayers.'

Will it help? Musk is a perfectionist with a tendency toward micromanagement. The Tesla Files contain internal tickets documenting his complaints about his car's navigation system. In one case, he claims the vehicle picked a subpar route. An employee records the incident dutifully. Soon, a group of engineers is assigned to teach Musk's car – and the entire Tesla fleet – the route he prefers.

One man's driving preference becomes a team-wide project – not because the system is broken, but because it doesn't match the boss's personal habits. That way, even airflow becomes a leadership issue.

In November 2016, chief designer Franz von Holzhausen writes to an engineer: 'Hey, sitting in Elon's new S and he noticed that the HVAC at start-up only blows warm air on your feet. Thinks that's odd.'

Musk had been in the car 'about one minute when I jumped in,' von Holzhausen reports, 'and that's when he started fussing with the controls.' After switching to manual mode and then back to automatic, the air-conditioning turned down again. 'Does this sound right? Elon said he noticed it in another car as well, and doesn't think the algo-rithm makes sense.'

The employee immediately creates a ticket. The title: 'Elon has concerns that heating in floor mode is not the right behavior for the Bay Area type conditions.'

And so, at Tesla, the quest to fulfil every single one of Musk's wishes continues. Some seem petty. Others, alarming.

In June 2020, Musk's employees are tasked with re-designing the structure of the Model Y – specifically, how its parts fit together. The goal: everything should interlock like Lego bricks, leaving no visible gaps.

In the relevant ticket, an engineer notes that he's work-ing on a solution for 'Elon's request to "lego fit" doors to body, with no oversized holes in the hinges allowed.' But the fix, he warns, 'unfortunately completely destroys our structural concept in this area.' In the comment section, he adds a final note: the change would cause a 'pretty significant drop off in performance' and result in 'non-acceptable performance.'

Still, the engineer reports that the production line – at Musk's request – is 'now pushing hard for this change.'

At Tesla, Musk's whims seem to override every argu-ment – even in matters of life and death.

During our reporting, we come across the issue of door handles. On Teslas, the handles retract into the doors while driving. The system depends on battery power. If an airbag deploys, the doors are supposed to unlock automatically, and the handles extend. That's what the Model S manual says.

The idea for the sleek, futuristic design stems from Musk. He insisted on retractable handles, despite repeated warnings from engineers. Today, Tesla's door handles are considered a potential safety hazard. Since 2018, they've been linked to at least four fatal accidents in Europe and the U.S.

Five people died.

By early 2024, our expertise on Tesla has become widely recognized – and we begin to gain access to more detailed information about cases like these.

In February, we report on a particularly tragic example: a fatal crash on a country road near Dobbrikow, in Brandenburg. Two 18-year-olds are killed when the Tesla they're in slams into a tree and catches fire. First responders can't open the doors – the handles are retracted.

The teens burn alive in the back seat.

A court-appointed expert from Dekra, one of Germany's leading testing authorities, later concludes that, given the retracted handles, 'it qualifies as a malfunction.' According to the report, 'the failure of the rear door handles to extend automatically must be considered a decisive factor.'

Had the system worked as intended, 'it is assumed that rescuers might have been able to extract the two backseat passengers before the fire developed further.'

Without what the report calls a 'failure of this safety function,' the teens might have survived.

Our investigation makes waves. The Kraftfahrt-Bundesamt, Germany's Federal Motor Transport Authority, gets involved and announces plans to coordinate with other regulatory bodies – including the UN Economic Commission – to revise international safety standards.

After our article appears in *Handelsblatt*, Germany's largest automobile club, ADAC, issues a public recommendation: Tesla drivers should carry emergency window hammers. In a statement, ADAC warns that retractable door handles can seriously hinder rescue efforts. Even trained emergency responders, it says, may struggle to reach trapped passengers.

Tesla shows no intention of changing the design.

That's Elon Musk. Because he prefers the sleek look of Teslas without handles, he accepts the risk to his customers. His thinking, it seems, goes something like this: at some point, the engineers will figure out a technical fix. The same logic applies to his grander vision: autonomous driving.

Because Musk wants to be first, he lets customers test his unfinished Autopilot system on public roads. It's a principle borrowed from the software world, where releasing apps in beta has long been standard practice. The more users, the more feedback. Over time – often years – something stable emerges. Revenue and market share arrive much earlier. The motto: If you wait, you lose.

Musk has taken that mindset to the road. The world is his lab. Everyone else is part of the experiment. It's a radical idea: treating a software crash the same way as a two-ton vehicle colliding with a pedestrian. To understand how Musk justifies that approach, we need to look at his philosophy.

Love of Humanity

In a TV interview in March 2012, Elon Musk was asked why he chases after goals that seem unobtainable.

'If something is important enough, you do it even if the odds are stacked against you,' he replied. He was referring to the founding of SpaceX. Musk had invested $100 million into his vision of conquering space, even though he estimated there was 'a ten percent chance of lasting in the long run.'

Since then, he's repeated those words – more or less verbatim – in countless settings. The quote is printed on posters, notebooks, and office decor. Tesla has elevated it to a kind of mission statement. It appears in internal presentations and onboarding materials. The message: anyone working at Tesla must push through any obstacle. No matter the cost. No matter the setbacks. In Musk's words: 'If things are not failing, you are not innovating enough.'

That quote has become a cornerstone of the philosophical framework behind Musk's empire. In a *New York Times* interview, he once called it 'a philosophy of curiosity.'

He said he was 12 when he entered an existential crisis. He began asking himself what the meaning of life was. Why go on living? 'I read the religious texts. I read the philosophy books,' he recalled. 'That – especially the German philosophy books – made me quite depressed, frankly. I recommend you not read Schopenhauer or Nietzsche as a teenager.'

Then he discovered *The Hitchhiker's Guide to the Galaxy* by Douglas Adams. That, Musk says, changed everything. The problem, he decided, wasn't that humans didn't have answers – it was that they didn't know what questions to ask.

'My life is finite – really a flash in the pan on a galactic timescale. But if we can expand the scope and scale of consciousness, then we are better able to figure out what questions to ask about the answer that is the universe,' Musk said. That, he believed, is how humanity might one day find its purpose.

Apparently, he has found his. Musk's highest ideal is freedom – first and foremost, his own. He belongs to a cohort of Silicon Valley entrepreneurs who champion the principles of libertarianism: the belief in the absolute freedom of the individual.

For liberals, freedom ends where another person's begins. Libertarians like Facebook's Mark Zuckerberg or investor Peter Thiel see it differently. More than that – they believe freedom and democracy may not be compatible at all.

Competition? Optional. Rules? A nuisance. Government? Best kept out of the way. What unites these thinkers is a conviction that regulation stifles progress – and that growth, if left alone, will manage itself.

Why trust bureaucracy when you have technology?

In the view of the Silicon Valley giants, society should simply let them build; the solutions will follow. Prosperity will rise. Problems will shrink. Yes, there may be collateral damage. But – and this is central to Musk's worldview – over the very long term, the bold visions of tech billionaires will be worth the cost.

In August 2015, Musk met the Scottish philosopher William MacAskill at a conference in California. The event was hosted by the Effective Altruism movement, which promotes doing the greatest good with the resources available. Musk spoke about the dangers of artificial intelligence. MacAskill used the moment to encourage billionaire investors to think ethically – and long-term.

The two stayed in touch.

In March 2022, as Musk publicly questioned whether the social media network Twitter still upheld free speech, MacAskill reached out with a suggestion: Sam

Bankman-Fried, the founder of the crypto exchange FTX, might be interested in buying the platform. Would Musk take the meeting?

He asked whether Bankman-Fried had enough money – and whether MacAskill could vouch for him. 'Very much so!' the philosopher replied. Bankman-Fried, he said, was 'very dedicated to making the long-term future of humanity go well.'

Later that same day – March 31, 2022 – MacAskill made the introduction. 'Hey, here's introducing you both, Sam and Elon,' he wrote in a group chat. 'You both have interests in games, making the very long-run future go well, and buying Twitter. So I think you'd have a good conversation!'

A deal never materialized. Bankman-Fried had doubts – and soon much bigger problems. By the end of 2022, FTX had collapsed. In March 2024, he was sentenced to 25 years in prison for fraud and other crimes.

It's unclear whether Musk resented MacAskill for vouching for a man who stole billions. But in August 2022, when MacAskill's new book *What We Owe the Future* hit shelves, Musk promoted it with a single line: 'Worth reading. This is a close match for my philosophy.'

MacAskill is a leading voice in a school of thought known as longtermism – a philosophy that's gained traction in Silicon Valley. Its core belief: every decision should be judged primarily by how it affects the distant future. Not just the next ten years, but the next thousand. Or ten thousand. Or more.

Critics argue that longtermism is just a dressed-up version of old clichés: You can't make an omelet without breaking eggs. The end justifies the means. What it often amounts to, they say, is a way of minimizing today's problems with vague promises of a better tomorrow.

At the heart of longtermism lies a simple premise: most people who will ever exist haven't been born yet.

A typical mammal species survives 1 to 2 million years. The oldest fossils of *Homo sapiens* are 300,000 years old. Even conservatively, that suggests humanity could have a million years still ahead. MacAskill estimates that 80 trillion people are yet to be born. For every one person alive today, 10,000 more may follow.

From that perspective, today's famines, disasters, and wars shrink into statistical noise. Even a 90 percent drop in the global population, MacAskill writes, wouldn't move the needle much in the long arc of human history. It's on that logic that thinkers — and tech titans like Elon Musk — base their cost-benefit calculations.

Take this example. According to the World Health Organization, 1.2 million people die in traffic accidents each year. In January 2017, Musk claimed that the next version of Tesla's Autopilot could reduce that number by 90 percent — roughly a million lives saved annually. Over 100,000 years, that would amount to 100 billion lives. Over a million years, a trillion humans saved. Even if a few million people were to die while testing the system, Musk's long-term ledger would still look brilliant.

What did he once tell his employees? 'If things are not failing, you are not innovating enough.'

Musk firmly believes that Autopilot will 'save lives' – in the long run. He said so in 2017, and he's said it many times since. Just as clear is the fact that, for him, the end justifies the means. For years, he has argued that his software improves fastest when real drivers use it in real traffic. Accidents become feedback. Fatal crashes become learning opportunities. Milestones on the road to a better future.

This is how Musk's logic turns risk into righteousness.

The Tesla CEO sees himself, then, as someone who loves humanity. A philanthropist. No matter how dangerous his products may be today, they are – in his view – blessings for tomorrow. In April 2022, at the TED technology conference, he called his companies 'institutions for the good of humankind.'

'SpaceX, Tesla, Neuralink, the Boring Company are philanthropy,' Musk said. 'If you say philanthropy is love of humanity, they are philanthropy.' Then he listed them one by one. Tesla accelerates sustainable energy. 'This is a love of philanthropy.' SpaceX aims to secure long-term human survival by colonizing other planets. 'This is love of humanity.' Neuralink will help heal brain injuries and reduce existential risks from AI. 'Love of humanity.' The Boring Company wants to fix traffic. 'If you care about the reality of goodness instead of the perception of it: philanthropy.'

CHAPTER 8

Ultra Hardcore

Elon Musk only wants the best for his employees. Or so he says.

In November 2023, the Tesla CEO visits his factory in Grünheide, just outside Berlin. He wears jeans, a Tesla shirt, and an open flight jacket. Employees later recall how effusively Musk praised the German plant that day. How he told them they were in good hands.

'A lot of times the quality of life at work is a bunch of small things,' Musk says. 'So it's like making sure that the way you get in is good, that the path to and from the factory is good, that the food is good, that it's a positive vibe in the factory. Like, I'm a big believer in playing music and just enjoying yourself.'

He knows full well that this isn't a given: 'When I was a young guy, unfortunately there were some places I worked where it was miserable. And I swore to myself, if I had a company one day, I want to make sure it's not miserable, and that people look forward to work. And that is what I want for you.'

Fast forward to September 2024. Some of the employees who cheered Musk a year earlier are now among our sources. Many are disillusioned. They recognize their own frustration in that voiced by Lukasz Krupski. They've read our portrait of the whistleblower, and say: That's my story, too.

They complain about the conditions inside the plant. About the pressure. About their supervisors. The speeches, they say, don't match the reality.

Nearly a year after Musk's monologue on what good work should look like, the mood inside Tesla's German factory has shifted. At a staff meeting, plant manager André Thierig and HR chief Erik Demmler take the stage and raise an issue that has been festering for weeks: unusually high numbers of employees off work due to illness – recently reaching 17 percent. Then they drop a bombshell that will dominate headlines for days to come: Tesla executives have been showing up unannounced at the homes of employees on sick leave.

'The level of absences was simply unacceptable,' says Demmler. 'So, we had to go to people's homes. And that's what we did. [. . .] We picked out 30 employees who showed certain patterns – people who had been off sick for long periods, but also those who'd submitted multiple first-time doctors' notes. And what we found was very, very mixed.'

One employee had just returned from the pharmacy 'with a huge bag full of painkillers. He didn't even need to say he was in pain – it was clear right away, as he opened the bag,' says the HR chief. 'So, we spoke with him. And he said: Hey, want to come in, have a coffee? I thought that was great. Unfortunately, we didn't have the time.'

Other employees 'really opened the door' as well, says Demmler. 'But there were also those where we definitely felt . . . The door opened and you thought, well, maybe they could – who knows? But clearly the will wasn't there.'

The executives didn't go out to reprimand anyone, Demmler insists. They simply wanted to ask: 'How are you doing? Can we help somehow?' But that intention often met resistance. 'You could feel it, the aggression,' says Demmler. 'Doors slammed in our faces. Threats to call the police. People asking why we hadn't made an appointment first.'

Our article revealing Tesla's new practice concerning sick workers becomes the most-read story of the week. Almost every major German outlet picks it up. Many consult labor law experts for legal assessments of what's been happening in Grünheide; others publish commentaries on the climate of fear inside Tesla's German plant. Television crews ask people on the street what they think of it all. Radio stations call to interview us about the story 'the whole country is talking about,' as one colleague puts it in an email.

On Musk's social platform X, the German satire magazine *Titanic* posts an image of the Terminator: a metallic skeleton with glowing red eyes holding an oversized machine gun. The caption reads: 'Tesla innovation. Home visits to sick employees now automated.'

Der Postillon, Germany's best-known satirical news site – comparable to The Onion – joins in. It reports that Tesla executives are now visiting the graves of deceased employees to check if they're really dead. We can't remember this ever happening to any other story in our paper.

Just as things begin to quiet down, Tesla CEO Musk comments on X: 'This sounds crazy. Looking into it.' A new wave of international media coverage begins.

Working for Elon Musk is no easy task. The Tesla CEO made that clear back in 2012, in an email to staff. At the time, the electric carmaker had just begun production. 'Please prepare yourself for a level of intensity that is greater than anything most of you have experienced before,' Musk wrote. 'Revolutionizing industries is not for the faint of heart, but nothing is more rewarding or exciting.' He promised to personally ensure that 'those who produce exceptional results are rewarded exceptionally, as is fair and right.'

This chapter is about what's left of that promise. After two years of reporting on Tesla, our conclusion is:

not much. In Musk's system, speed comes first, dissent is not an option. Employees pay the price. Tesla exposes them to safety risks, underpays them, monitors them. Pressure is a constant. Nowhere is that more visible than at the Grünheide plant – where Musk's methods collide with German ideas of what makes a good employer. The unannounced home visits struck a nerve because they perfectly capture Tesla's toxic workplace culture. Musk himself has a name for it: *'ultra hardcore.'*

The turnover at Tesla is ultra hardcore, too. Whistle-blower Lukasz Krupski casually mentioned early on that Tesla 'recycled' 20 percent of its staff each year. We weren't sure whether that could really be true. But with the help of the 'Termination Master Report' – an internal spreadsheet from the leaked Tesla Files – we find our answer.

The Excel file lists nearly 200,000 employees who left Tesla between 2007 and early 2022. We compare the number of departures each year with the size of the work-force at the time. At first, we can't believe the result. We run the numbers again and again – and keep arriving at the same figures: Tesla's turnover rate is more than 30 times higher than that of comparable companies. In 2018 alone, about 70 percent of Tesla employees left the company. The following year, the rate was similarly high, then it began to decline. In 2021, Tesla still had to replace 40 percent of its workforce. For comparison: Volkswagen reports an annual turnover rate of just 1 percent.

Why is the rate so much higher at Volkswagen's American competitor? The 'Termination Master Report' includes the reasons employees left. Of the 44,000 workers who exited Tesla in 2021, a third were let go. In 2020, there were roughly 37,000 departures – more than 15,000 of them initiated by the company.

We spend hours reviewing the file. Some dismissal cases are strange – and occasionally absurd. In 2021, Tesla fired an installer in California after he urinated in a customer's backyard. The comment field notes: 'Admitted to management he has done this before. No extenuating circumstances.'

Other employees were let go for playing chess on their iPhones at work, smoking marijuana on factory grounds, or assaulting supervisors. One was fired for using Tesla-paid Uber rides to visit private destinations nearly 60 times. Another lost his job because his laptop bore a sticker that read 'MILF Hunter.' MILF stands for Mother I'd Like to Fuck.

Most dismissals, however, are less absurd. Tesla typically cited restructuring, chronic absenteeism, or poor performance. When employees leave voluntarily, they often blame bad management or low pay.

The list of criticisms we gather from insiders is longer – and far more detailed. Each interview adds texture to the cold, mechanical figures in the report. Over time, we hear more and more strange and sometimes surreal anecdotes about life at Tesla. And gradually, a pattern begins to emerge.

Tesla's House-Elves

Why Tesla? It's the first question we ask every new source – and the answers sound strikingly similar. Whether they still work there or have moved on, all say they joined the American company out of ideological conviction or admiration for Elon Musk. Some were driven by a desire to find meaning in their work and help make the world a better place. Others describe joining out of sheer reverence for Musk – the exceptional entrepreneur who makes legacy automakers like BMW and Volkswagen look like relics from another age.

Our whistleblower Lukasz, too, once called himself a Musk disciple. Tesla's leadership welcomes candidates like him with open arms – and does plenty to feed those expectations. They speak of 'our mission' and 'Elon's master plan.' That works. Many tell us they joined Tesla not just to make money, but to become part of something bigger than themselves – a company that doesn't just build cars, but moves history forward. The company's code of ethics frames it like this: 'Tesla has been, is and always aspires to be a Do the Right Thing company – in other words, engaging in conduct that you and your family would be proud of,' it reads. 'Tesla's mission to accelerate the world's transition to sustainable energy is itself grounded in Doing the Right Thing.'

Those invited to Tesla job interviews describe nearly identical experiences. They were surprised to find that their qualifications mattered less than their motivation. 'They didn't ask what I could do. They just wanted to know if I was excited about Tesla and the job,' recalls one employee at the Gigafactory in Grünheide. He had the impression that recruiters were mainly looking to weed out people 'who check the clock too often.'

A former sales employee had a similar takeaway. In his interviews, the focus was 'hardly on my resume' but all about 'the right mindset.' Recruiters tested how strongly he identified with the brand. 'They asked why I wanted to work for Tesla, what the company's mission was, and how I thought I could contribute to it.' They also asked about his ability to handle pressure: could he 'cope with constant stress'? It was made clear that employees were expected to 'go the extra mile.' Those unwilling to do so, he was told, had no place at Tesla.

Several workers say they were eager to give their all. Some even accepted lower pay just to get a foot in the door. Compared to German competitors, Tesla pays poorly. Over time, they say, this is unsustainable. The Tesla shine wears off – and with it, the power to attract or retain top talent. For many, the job eventually becomes simply 'not worth it.'

Anyone with a mortgage and children, says one employee, will sooner or later start asking: 'What's the mission doing for me?' Another recalls how he and his colleagues often wondered what they were still 'pushing so hard' for. He's convinced that any mid-sized company offering 'a few euros more' could pull up a bus to the gates of the Grünheide plant and 'poach technicians by the dozen.'

Even Tesla's works council admits there's 'room for improvement' in remuneration. At the top of the hierarchy, however, is Elon Musk. Negotiations with management, a council member remarked at a 2023 staff assembly, 'aren't as simple as you might think.'

Some employees stay loyal to Musk and his vision for years. Others come to a different conclusion much sooner. They realize that Tesla is not the dream employer they had imagined – not the mission-driven, forward-looking company they thought they were joining. Over time, the relentless pressure takes its toll. 'At Tesla,' one insider tells us, 'it's always about growth, growth, growth.'

We're not surprised. By now, we've reviewed the company's internal training materials – documents that lay bare the mindset Tesla seeks to instill. They, too, are part of the Tesla Files.

One internal presentation, titled 'Growth Mindset for Leaders', calls on managers to set 'ambitious goals,' take risks, and treat mistakes as learning opportunities. The core principle is simple: success is earned through effort. 'In a nutshell,' the document explains, 'people with growth mindsets believe that they can get smarter, more intelligent, and more talented through putting in time and effort.'

But Tesla doesn't just want this mindset at the top. Managers are told to imprint it on their teams. In its training materials, Tesla references the Pygmalion effect – the psychological phenomenon where people rise or fall to meet expectations. Managers are urged: 'Leverage the power of positive expectations and bring your team's potential to life!!'

To show them how, Tesla provides a development plan. According to the company's own HR estimates, only 10 percent of employee development stems from training and formal education. Twenty percent comes from mentoring and 'informal feedback.' The remaining 70 percent? 'Experiential learning.'

What does that mean in practice? It means employees are expected to grow by taking on 'stretch assignments' – projects beyond their current skill set that force them to adapt, perform, and deliver. It means volunteering for 'extra responsibility,' even when workloads are already high. The goal is constant acceleration – always more, always faster.

But what Tesla frames as opportunity, many employees experience as overload. The company religiously pushes its people past their limits. It often leads to physical and mental strain. But at Tesla, this seems to be taken not as an unfortunate side effect, but something to value, even encourage. For a company that prides itself on innovation, its people strategy is built on an old formula: pressure creates diamonds.

At Tesla, pressure is constant. And those who don't fall in line face consequences.

Employees are evaluated twice a year using a five-point scale, from 'unacceptable/poor' to 'exceptional/outstanding/highly satisfactory.' Performance alone isn't enough. Supervisors must also assess how well employees embody the company's values – whether they are 'committed to Tesla's mission' and willing to 'do what's right, even when it's hard.'

A low score sets off a formal process: the performance improvement plan. Managers must document exactly where the employee is falling short – and by when they are expected to improve. A presentation titled 'Performance Acceleration' lists the potential outcomes: 'possible employment actions such as a demotion or termination.'

Tesla's hire-and-fire culture doesn't stop at the lower levels. The Tesla Files include internal evaluations in which managers rate other managers. One spreadsheet, filled out by Andrew Baglino – Tesla's longtime top engineer for powertrains and energy – scores colleagues on categories like 'driven,' 'collaborative,' and 'trustworthy.'

Baglino praises one leader for setting 'aggressive goals' to motivate their team, advises another to 'avoid being defensive,' and urges a third to apply 'a hard push now and again to drive the sense of urgency.'

Almost empathetic by contrast are Tesla's instructions for how to conduct a termination meeting. In a guide titled 'How to Master Challenging Conversations,' the company offers tips for difficult moments – whether it's giving negative feedback or informing a team of a tough company decision. Tesla lays out exactly how these conversations should be orchestrated. Ideally, managers should arrange 'a suitable meeting room (no open space or glass wall) in which you won't be overheard,' be 'compassionate,' but 'not sugarcoat the message.' Timing, too, is key: 'Rather have those talks at the beginning of the week, late in the day and not on sensitive dates (e.g. birthday).'

When we read these passages to former employees, some laugh out loud. Their own experiences were very different. Several say they discovered they'd been fired when they couldn't log in to their laptops, or when their badges suddenly stopped working at the factory gate. A former technician from Grünheide tells us he once found a calendar entry in Outlook titled 'Promotion decision' – only to be informed of his dismissal by people he had never met.

When he entered his office, his bag was already packed by the door. He was told to hand over his company badge. Two security guards flanked him the entire time. He wasn't even allowed to use the restroom. If he needed to go, they said, he could relieve himself outside – even though it was December and temperatures in Grünheide were below freezing. The guards escorted him until the last turnstile clicked shut behind him. Then they turned and walked away without a word.

Stories like his multiplied after April 16, 2024. That day, we were the first outlet to report on an internal email from Elon Musk to his staff. The subject line: 'Reorg.' Tesla had made the 'difficult decision' to reduce its 'global headcount by more than ten percent.' Musk wrote, 'There is nothing I hate more, but it must be done. This will enable us to be lean, innovative and hungry for the next growth phase cycle.'

Many of those affected took to LinkedIn. One software engineer, who had worked at Tesla for more than five years, shared his layoff post alongside a picture of Dobby, the house-elf in the *Harry Potter* films – a character known for loyalty and self-sacrifice, even to the point of masochism. The caption read: 'Dobby is free.'

'That image perfectly sums it up!' another laid-off Tesla employee comments beneath the post. The engineer responds with a thumbs-up, followed by a quote from the *Harry Potter* novels describing the house-elves – a comparison he clearly sees as fitting for Tesla workers, too: 'House-elves are known for their unwavering loyalty to their masters or mistresses. They take great pride in serving their households and often prioritize their master's well-being above their own.'

The Safety Illusion

Tesla employees put their safety in Elon Musk's hands. 'Each Tesla employee has the right to return home safely to his or her family each day,' reads the company's code of ethics. The goal, it says, is to keep the number of injuries as low as possible. Tesla makes a promise to its workforce: 'We are committed to safeguarding this right and will not compromise it for production or profit.'

And yet, Tesla is notorious for its unusually high number of workplace accidents. Musk himself once described the company's 'production hell' as the eighth of nine levels. The Tesla CEO was borrowing from 'Inferno', the fourteenth-century vision of Hell by the Italian poet Dante Alighieri. In it, Dante descends through nine circles of suffering, each more punishing than the last. The final circle is where Satan waits.

Many employees still see *Hell* as a fitting metaphor for the environment inside the factory: chaotic, exhausting, and riddled with risk. Production targets take priority. Safety, they say, too often comes second.

A quick search through the Tesla Files for the terms

'injury' or 'workplace safety' paints a vivid picture. We find internal logs documenting work-related injuries and illnesses in Tesla's U.S. factories – the same reports the company is required to file with the Occupational Safety and Health Administration, the federal agency tasked with overseeing workplace safety.

The list is grim. Body parts crushed in vehicle frames. Skin burned by chemicals. A worker struck in the back by a falling box from a forklift. Another breaks his nose after being hit by a moving pallet.

There's also a list of incidents and corrective actions intended to prevent future accidents – more than 5,000 entries in total, and just as revealing. One states: 'I will have maintenance supervisor look at the conveyor for any excessive wear that could have caused the module to fall from the conveyor.' Another: 'Next time stop the forklift and make sure no one or nothing is in the way or in danger. Needs to be re-certified.' A third notes: 'Associate need to understand that being highly motivated and dedicated is great but not at the risk of injury. Associate needs to understand that once the help has arrived her primary job is to supervise the work and not necessarily perform the work herself. This has been communicated to her.'

Insiders from Tesla's U.S. factories describe 'intolerable conditions.' They speak of waste piles, unstable shelving, and punch clocks installed in dangerous spots – like truck-loading zones. One worker even provides photographic proof. Others recall an incident at the Fremont plant when a sewage pipe burst and workers were told to keep going, to literally wade through shit. Dennis Duran, who worked in the paint shop, told Bloomberg what happened when some refused: they were told, 'Just walk through it. We have to keep the line going.'

There are similar stories from Germany. While speaking with workers at the Grünheide plant, someone shares a letter sent to the national wood- and metalworkers' compensation board. A Tesla employee complains about safety conditions inside the factory. 'It took us decades to win real protections for workers,' he writes. 'Tesla is actively sabotaging them.' He claims to have proof that electrical installations on 'at least two separate production lines' are fitted with improper CE certifications. We're able to review the corresponding photos. The CE mark is a certification indicating that a product meets EU standards for safety, health, and environmental protection.

The letter also alleges that Tesla is violating regulations on airbag installation. Airbags are classified as pyrotechnic devices, and handling them requires proper training. At Tesla, this often isn't the case, the employee claims: 'Production numbers matter more than sending employees to vital safety trainings.' Equipment is also said to be lacking. Despite 'constant loud noise,' there are not enough earmuffs to go around. Sanitary conditions are so poor, he writes, that he often can't even wash his hands before breaks. He closes his message with a warning: 'We cannot allow American conditions in Germany.'

The German news magazine *Stern*, which was investigating Tesla at the same time we were, confirmed our findings. One detail from their report we wish we had discovered ourselves: in its very first year of operation, the Grünheide plant requested an ambulance or helicopter 247 times.

In the summer of 2023, the head of the workplace safety team addressed the issue on stage. 'We've had a lot of workplace accidents,' he said. 'Fortunately,' he could report, the rate was falling – at least in relation to the total number of hours worked. Still, it remained crucial to 'reduce risks gradually.' His team now aimed to 'implement at least one measure per accident' to ensure 'that the same accident doesn't happen again.' To that end, employees were urged to report every incident. 'So we can really take a close look at what to do,' he said, 'ideally together with their supervisor.'

But not all safety concerns involve machinery or workflow. A very different kind of danger stems from the sexism and racism that employees repeatedly report. The Tesla Files show that in 2020 and 2021, the company recorded a total of 80 dismissals for 'harassment.' The data doesn't specify the nature of the offenses. But it's clear that sexism exists. A May 2022 works council meeting at the Grünheide factory includes this agenda item: 'Harassment of female Tesla employees by third-party contractors — How can we as a council protect and support the women?'

Former U.S. employees also describe widespread discrimination. At the Fremont plant — on the assembly line, in restrooms, and in offices — an 'openly racist environment' was reportedly the norm. Swastikas had been scrawled on bathroom walls; the Tesla Files include records of such incidents. In 2021, the company fired an employee for spray-painting swastikas onto cast parts. That same year, an American named Owen Diaz sued Tesla, claiming the company failed to act against racism in Fremont. A California court awarded Diaz $3.2 million in damages for 'emotional distress.'

And then we come across a source of conflict in the Tesla Files that truly surprises us: the company is afraid of its own customers.

Tesla's security team prepared a six-page internal guide for sales managers titled 'Threats and Aggression Toward Your Team.' It outlines scenarios in which employees may face 'verbal harassment, threats or even physical aggression from outside groups.' The document notes that the security team had already received reports of 'truly awful situations.' Customers had threatened to harm staff, demanded their personal information, or sexually harassed them.

To help prepare teams, the guide recommends that managers organize 'something like a monthly coffee session where everyone in your team can openly share details.'

Employees should learn to 'give customers time and space to have their say.' Verbal abuse, the document advises, shouldn't be taken personally: 'The aggressive individual is likely venting at the nearest available person, not you personally.'

Managers are encouraged to report any interaction that made them 'uncomfortable or concerned.' But the security team is also clear: 'Unhappy interactions are part of the job.'

The document goes on: 'At some locations,' the document states, 'managers have overseen construction changes to the reception area that provides employees with an immediate exit to a separate room in the event that someone becomes violent.' In other words: at Tesla, some employees now have access to what could be called panic rooms.

The Machinery of Control

Tesla does everything it can to ensure that details like these never reach beyond the factory walls. The company seals itself off, both physically and reputationally. Even before job interviews, applicants are asked to sign a non-disclosure agreement. 'Tesla and Applicant wish to evaluate a possible employment or independent contractor relationship ('Purpose'),' the document begins.

To that end, Tesla may share confidential information – information the applicant is obligated to 'protect [. . .] on the terms set forth below.' Confidential information is defined broadly: anything marked as such, verbally stated, or that the applicant 'should reasonably understand to be confidential or proprietary.'

Signing an employment contract means pledging silence a second time. 'Any discussion of internal matters may have a material impact on Tesla, affect our reputation, and expose us legally,' states the company's communication policy. 'Please remember when communicating with others that you may be held personally liable or subject to prosecution if you violate the confidentiality agreement.'

Tesla sums up the rules in an internal document titled 'How We Present Tesla to the World.' Employees are not allowed to speak for the company. And anyone who leaks internal information, the document warns, may 'feel important and "in the know" – but it will also make you unemployed.'

Prevention is one thing. Control is another. Employees tell us about 'Tesla, the surveillance state.' The company doesn't just silence them with strict NDAs – it cultivates a climate of fear. Surveillance is ever-present, insiders say. Criticism is shared only in private, if at all.

'Most people are afraid of pushback or consequences if they speak up,' says one employee in Grünheide. 'At

Tesla, you always have to watch who you talk to – and what you say. Otherwise, your supervisor might come down on you.'

Where no supervisor can look over your shoulder, Tesla turns to digital tools.

In the case of Lukasz Krupski, we discovered that Tesla uses software from U.S. company Code42, which specializes in identifying employees who may pose 'a threat' to their employer. The programs monitor who moves which data, and where. 'Not all unusual behavior will be problematic – but they should be investigated regardless,' the company's website says, promoting its insider-threat detection products. Elsewhere: 'You need to detect and respond before they walk out the door.'

The Tesla Files contain numerous Code42 invoices. Between November 2020 and May 2021 alone, Tesla purchased 31,000 licenses for the surveillance software. One invoice from November 2020 totals $826,000. Including support and 4,000 additional licenses, the bill exceeded $1 million.

All that money didn't stop Lukasz from extracting internal data, copying it, and ultimately passing it to us. Not exactly a glowing review for the product. And yet, Code42 now uses the Krupski case as a sales pitch.

On its website, the spyware company lists '11 Real-Life Insider Threat Examples.' The very first: the former Tesla employee who leaked 'sensitive personal data to a foreign media outlet.' Code42 concludes: 'The stain on the brand's security reputation is irreversible.'

When information leaks, Tesla wastes no time going after those who break the code of silence. The pursuit is carried out with both professionalism and fervor. The carmaker employs former intelligence operatives – as revealed by the Krupski case and by bills and documents in the Tesla Files. For years, Tesla contracted two U.S. security firms: Nisos and Gavin de Becker, both of which employ former CIA and NSA agents, as well as veterans of the military, FBI, and other federal departments.

According to its own materials, Gavin de Becker serves 'more than 80 of the world's most prominent families and at-risk individuals.' The firm offers threat assessments, investigations, and background checks. Its team primarily provided private security for Elon Musk and his family. But in September 2014, its Threat Assessment and Management Division also billed Tesla for 'investigative research' and a covert operation in San Francisco. What exactly happened remains unclear.

Nisos, for its part, billed Tesla nearly $1 million in July 2018. The invoice included charges for assistance with a 'leak investigation' and 'threat hunting and attribution.' The timing was notable: just days earlier, Tesla had sued engineer Martin Tripp for $167 million after he tipped off reporters about alleged production delays.

'Detecting, Disrupting, and Preventing Adversary Operations' – that's how Nisos describes its work. The goal of its intelligence-driven investigations is to provide clients with the insights needed to 'stop your adversaries.' One tagline reads: 'Stop playing whack-a-mole and get to the root cause.'

Today, Tesla appears to rely less on Nisos and Gavin de Becker. Instead, the company now directly hires former intelligence officers to monitor employees and plug internal leaks. The unit responsible is called Security Intelligence Investigations – and it has gone after our sources too, as we saw firsthand through Lukasz's legal battle with Tesla.

The company's intelligence unit doesn't confine itself to Tesla property. In late 2022, Tesla posted a job ad for its German factory. It was looking for a 'Security Intelligence Investigator' with experience in law enforcement or intelligence work – someone to be tasked with gathering information about employees 'both inside and outside Tesla's walls.'

Two years later, Tesla expanded its security team again. A new posting read: 'In this role, you will conduct both proactive and reactive investigations and actively respond to internal and external threats to Tesla's protected and confidential information.'

Just days earlier, we had published another article with internal details from the Gigafactory in Grünheide – one that, we're told, annoyed Tesla management. The company's renewed search for informants may, of course, be a coincidence. Then again, maybe not.

Cracks in the System

It took many months, but today we can say Tesla's wall of threats and confidentiality clauses no longer stops us. In the first half of 2023, we struggled to get anyone from the Tesla world to speak with us. Now, we can verify even sensitive information from multiple sources within hours. Tesla's defensive tactics sometimes remind us of a cartoon character standing in front of a leaking wall, frantically trying to plug one hole while water bursts through another. It's a hopeless task.

Two years after we published our first Tesla investigation, information is flowing our way from inside the company with ease. We got to know many of our sources well even before that first story ran. Some reached out after we published a call for tips on the *Handelsblatt* website. Others responded to our ongoing reports. Most are former employees who spent varying amounts of time at Tesla. But current employees also contact us – many to criticize working conditions or company culture.

Still others stand firmly behind Tesla, but are frustrated with how little the company communicates. So, they take it upon themselves to do just that. While some call their time at Tesla the worst job experience of their lives, others feel the opposite. 'If you haven't worked there, you can't understand – it's more than just a job, you live it' wrote one person who had recently been let go. 'From day one, it was the best thing that ever happened to me.'

Often, people like this assume we won't be interested in their stories because they have nothing negative to say. But we value every conversation. Every contact is another doorway to the truth. Another chance to confirm details, cross-check facts. Back when we began investigating Tesla, none of us imagined we'd one day have access to a whole network of sources like this.

What we hear from them is always revealing, often disturbing – and sometimes just absurd. In July 2024, an employee in Grünheide sends us an audio recording of a staff meeting. Our minds boggle.

The meeting begins with a heated argument over how employees are expected to call in sick. Factions within the works council attack one another. Union reps accuse management of harassment and ignoring workplace injuries; the head of the council fires back, calling them puppets of IG Metall headquarters – Germany's largest industrial union.

Then factory director André Thierig steps up to the microphone. The Tesla staff is in for a surprise. Their boss doesn't address injuries or harassment. His mind is on coffee mugs. Tesla can't seem to keep track of them.

'You all expect that the cupboard is just magically stocked every day with clean, fresh coffee mugs,' Thierig says. 'Let me give you a number. Since production started here, we've purchased 65,000 coffee mugs. Sixty-five thousand! Statistically speaking, each of you has five IKEA mugs at home.'

Thierig's tirade puts us in a moral dilemma. We know that a Tesla boss complaining about the theft of mugs has meme potential. On the other hand, isn't that too ridiculous? We are *Handelsblatt*. We don't want to dilute our credibility as a serious business publication by going for some extra clicks on our website.

So, we compromise. 'Audio Recording Reveals Internal Power Struggles at Tesla Works Council,' reads the headline of our article on July 10, 2024. For 12 paragraphs, we describe the internal conflict within Tesla's employee works council. Then we devote two paragraphs to the mug drama. Finally, we return to the real topic: the battle for power within the works council.

It takes less than 60 minutes for the story to blow up. Our article goes online at 9:10 a.m. By 9:47, *Spiegel Online* has published: 'Elon Musk Has Lost His Mugs.' Soon, new headlines appear every few minutes.

Media outlets outdo each other, trying to capture the absurdity of it all in a single line: 'Dish Disappearance,' 'The Mystery of the Missing Mugs,' and 'Tassla-Gate,' as one German outlet wryly put it. The story spreads like wildfire. The next day, it reaches the English-language media. *Fortune* reports on it, as does *The Business Standard*, along with countless Tesla blogs and fan pages. Days later, more outlets around the world pick it up. Even Musk himself joins in. On X, he writes: 'We were mugged!'

On a Thursday afternoon in mid-September 2024, we receive a message from a Grünheide source: 'André Thierig just gave *Handelsblatt* a shoutout.' That catches our attention. Over the past few months, we've reported extensively on Thierig and his speeches to employees. 'Is there another meeting today?' we ask. Silence.

We stay calm. By now, we know that reception inside the Grünheide plant is spotty, and employees often rely on the internal network to contact us. Many are too afraid to speak out from work, fearing Tesla might be monitoring them, so they only reach out from home.

That evening, we receive a download link to a new audio file – two hours long. It's a recording of the assembly where Tesla's local leadership addresses the controversial home visits to employees on sick leave.

We'd heard whispers about high absenteeism rates at the Grünheide Gigafactory, but now we have the numbers. According to the presentation by factory director Thierig and HR chief Erik Demmler, absenteeism reached 15 percent in early August and climbed to 17 percent the following week.

To put this into perspective, we compare Tesla's absenteeism rate with national and industry averages. According to Germany's Federal Statistical Office, the national average in 2023 was 6.1 percent. In the automotive industry, the rate was lower, at 5.2 percent, according to health insurer DAK. These numbers make Tesla's absence rate seem absurdly high.

The strange thing is: Thierig and Demmler apparently talk about all of this knowing full well every word will reach us. Over the past months, insiders have repeatedly sent us confidential emails, screenshots from internal chats, and audio recordings like this one. Indeed, as Thierig steps up to the stage and asks for quiet, he says, 'It would be great if everyone who's allowed to listen is actually listening – not just *Handelsblatt*, but especially everyone here in the room.' Then the German Tesla boss starts talking about his trips to employees who called in sick.

The next day, back in our office, we wonder: What exactly are Thierig and Demmler hoping to achieve? Sure, employer visits to employees on sick leave are not illegal – even unannounced. But whether legal or not, we're certain the sections about these sick-leave checks will make waves when we report on them – even more so than the coffee mug saga.

Don't the Tesla managers see this coming? Or do they want to spark a public debate about 'slackers in Germany'? There's a third possibility: perhaps these home visits are simply intended to intimidate the workforce. *We know where you live!*

Only Thierig knows, but he's not telling us. Our article, it seems, has put him in an awkward spot. Countless newspapers are now covering the sick-employee checks. The term 'Krankenpatrouille' has been coined – best translated as 'sick patrol.' But while Thierig can see the criticism increasing, he can never forget who he works for.

Elon Musk has never been shy about his expectations. His 'ultra hardcore' emails are legendary, as is his notorious habit of sleeping on the factory floor. And now, in Germany, nearly one in five employees are calling in sick? How is Thierig supposed to explain this to him?

Our story spreads across the globe. A colleague sends us a video from South Korea: two influencers sitting on a beach in Busan, laughing for 20 minutes straight about Tesla's home visits. Amusement is one common reaction. The other: suspicion. Everywhere, headlines announce that something is deeply wrong inside Tesla's German factory. Eventually, Thierig speaks to Germany's DPA news agency to defend the practice, calling it 'a standard measure.'

Not many share his view. The union IG Metall calls the home visits 'a bizarre move.' The real issue, they say, is this: when staff are missing, Tesla pressures the sick and overloads the healthy. If the factory leadership truly wants to reduce absenteeism, 'they need to break the vicious cycle.'

Of course, everyone knows who sits at the top of Tesla: Elon Musk. Christian Görke, a left-wing MP and parliamentary director in the German Bundestag, comments: 'The richest man in the world needs to realize this isn't the Wild West.' Even German billionaire Carsten Maschmeyer – best known from the country's version of *Dragons' Den* – weighs in. Tesla's behavior is 'appalling,' he says. 'Employer home visits don't help anyone. [. . .] It's a troubling intrusion into employees' privacy.'

Tesla staff are discussing it, too. We receive screenshots from Teams channels inside the company. In them, employees vent about the stress of the job. 'Many Tesla employees – including myself – are struggling with health issues like exhaustion and sleep deprivation,' one writes. The cause, she says, is the shift schedule. A night shift followed by an early shift. Workers show up Monday morning after barely sleeping all weekend. 'I fall asleep behind the wheel on my way home,' she writes. 'It scares me. I'm writing this to get the works council's attention. This is serious – it's about our health.'

Dozens of colleagues respond with messages of support. 'Well said,' writes one. 'I'm personally not struggling [. . .] but I'd also prefer a night-late-early rotation.' Another adds: 'It's true. We need rest after work. Let's change this together.'

Neither Thierig nor HR chief Demmler addressed the shift system in the assembly – or do so in the public discussion that follows. Focusing on absenteeism, they assure employees that there's no 'blanket suspicion' being cast,

Thierig tells DPA that the numbers suggest 'the German welfare system is being exploited to some extent.' He claims he and Demmler were trying to appeal to employees' work ethic.

Thierig notes some 'improvement' but doesn't rule out further home visits. One clarification is important to him: Tesla's absenteeism does not indicate overwork. 'Our attendance analysis shows a clear pattern: about five percent more employees call in sick on Fridays and during late shifts. That's not an indicator of poor working conditions – because those conditions are the same across all days and all shifts.'

Thierig's home visits didn't come out of nowhere. Back in July 2023, when we received our first leaked audio, he was already railing against 'freeloaders' and 'late sleepers' in a staff meeting. 'We're not going to tolerate this anymore – some people breaking their backs for others who just can't be bothered to show up,' he said. In his factory, there's no room for people who 'can't get out of bed in the morning.' Tesla, he warned, wouldn't tolerate employees 'playing sick.'

Union Wars

A year later, Thierig changes course. The factory director announces that Tesla will begin rewarding employees in Grünheide who distinguish themselves by rarely calling in sick. Employees will be able to achieve various 'status levels' based on continuous attendance. Those who reach 'Gold status' will receive a €1,000 bonus. 'To be honest, we struggled with the idea,' says Thierig. 'Why should you reward someone just for showing up?' However, he adds, there are people in the factory who 'come in tirelessly and have incredibly low absences.' He says he's glad to see those efforts finally recognized.

In September 2024, the tone shifts once again. 'It's not honorable to exploit the system,' says HR chief Erik Demmler at the staff assembly where the home visits are discussed. Tesla, he explains, has carefully studied when absences occur. 'Why are Fridays always critical? What's with the late shifts?' Then he arrives at the final slide in his presentation. 'This analysis is very, very important. Because now we can examine these patterns – and figure out who we need to sit down with.' Of course, he adds, termination should 'only ever be the last resort.' But: 'It's unfortunate, but apparently that's what it takes for the message to get through. And to stop it from being exploited again.'

Tesla's leadership repeatedly stresses that the situation with the factory's 1,500 temp workers is entirely different. They work under the same conditions as everyone else – but have a sick leave rate of just 2 percent. Thierig announces new measures. At the end of his speech, he adds: 'And maybe one last message to *Handelsblatt*. If you're looking for a headline, just write: Tesla in talks with works council on early conversion of temporary workers.'

Sure enough, just weeks after Thierig's outburst over absenteeism, Tesla converts 500 temporary workers to permanent contracts. The company denies any connection between that decision and the pressure campaign on employees on sick leave. Instead, Tesla praises itself for giving 500 people a 'long-term perspective.' But after management publicly lauded the contractors' near-perfect attendance, many permanent employees see the move as a message: if you miss too much work, you will be replaced.

Tesla is also quick to clarify that IG Metall was 'not involved at any point' in these decisions. That part of the sentence is underlined in the company's press release. The power struggle between the union and Tesla's management has been simmering for a long time. Tesla is doing everything it can to keep IG Metall's influence in the factory as small as possible.

The anti-union stance comes from the very top.

'Well, I think we want to be, you know, in control of our own destiny. And if there is a sort of a third party that comes in, then they will be really in control. And it will not be a good situation,' Elon Musk says during a visit to the German Gigafactory in November 2023. 'What's your view on unions?' asks Thierig. Musk laughs. Then: 'You mean IG Metall? Say it. Yeah, no – I think the problem is we'd lose control of our destiny.'

Musk takes his anti-union stand one step further. 'In their case, their loyalties are divided amongst – are really primarily with – the other car companies. Are they actually going to make the right decisions for Tesla? It's very unlikely because their membership is overwhelmingly with other car companies. So they will favor those other car companies instead of Tesla. And I think it will be quite bad for us actually. So, it would kill the vibe, that's for sure.'

When the workforce elects a new works council in March 2024, IG Metall's union list wins 16 seats, making it the strongest single group. Four other lists, including that of the incumbent council chair Michaela Schmitz, claim 23 seats between them. Internally, these councilors – seen by many as close to management – are referred to as 'Faction 23.' They form a bloc against IG Metall.

Whenever possible, one side tries to outmaneuver the other. Workers are fed up. 'All this finger-pointing – it's annoying,' writes one person in Tesla's internal chat system. 'Right now, the works council is doing nothing for the employees.' Another complains: 'It's kindergarten with a whole lot of boo-hoo-hoo.'

The power struggle escalates in July 2024. Tesla fires a production worker who sits on the works council as part of IG Metall's slate. At the time, the shift supervisor is just four weeks away from starting parental leave – giving him dual legal protections. German labor law grants council members special protection against dismissal. Barring company shutdowns, they can only be fired for serious reasons.

The employee says Tesla demanded he withdraw his request for part-time parental leave – and threatened him with a formal warning and dismissal if he didn't comply. HR told him part-time 'doesn't work' at Tesla. Full-time only. He refused to sign the termination agreement. Then Tesla accused him of entering false working hours on two occasions. According to the employee's court testimony, these entries had been agreed with his superiors. It's common practice at Tesla, he says, to work beyond the legal ten-hour limit – and to compensate for it later. 'Everyone knows this.'

Tesla's move backfires. On September 23, the labor authority blocks the dismissal. 'The reasons cited in the request, regarding the employer's claimed repeated issues with working hours, are difficult to follow,' reads the ruling. 'There are inconsistent rules for working hours and time-keeping on the employer's side, which are not reflected in practice and in some cases result in legally impermissible outcomes.'

Then, in October 2024, Tesla fires another union-affiliated council member. Again, a production worker. According to him, on October 1, just after returning home from a night shift, he notices an email sent at 5:35 a.m. It's a last-minute invitation to a council meeting at 9 a.m. Exhausted and frustrated, but taking his role seriously, he returns to the plant.

At the end of the meeting, the agenda turns to feedback. The topic: how meeting invites are issued. Attendees voice their misgivings, complaining that short-notice invites are unfair, especially to shift workers, who don't have enough time to respond.

Finally, the overtired union rep speaks up: 'What the hell gives you the right to email me at 5:35 after a night shift and expect me here by 9? I was home. I had to come back. I don't know the legal situation, but if I find a loophole, I'll rip you a new one.'

Witnesses say the meeting chair seemed pleased with his outburst. Someone from Faction 23 reportedly remarked: 'Quick, write that down.'

The worker is expelled from the room and suspended. He immediately apologizes, clarifying that he meant legal – not physical – action. Still, Tesla fires him six days later, citing a threat of violence. The works council must approve the dismissal – and with Faction 23's votes, it does, just as it did in July. In Germany, it is almost unheard of for a council to vote out one of its own.

Union reps say they don't understand how Faction 23 can be so thin-skinned. Its members, they claim, have called union colleagues 'clowns' and told them to crawl back to the union house and 'go finger yourselves.' We can't verify that. But we do have audio recordings in which factory chief Thierig accuses IG Metall reps of not having 'the balls' to speak up during an assembly. Meanwhile, the reps in question were actually present in the room – but their request to take the stage had been denied.

IG Metall suspects a larger strategy. Of the 35 people on its list, 25 have received official warnings this year. Every single IG Metall councilor, they say, has been threatened with dismissal. The union believes this is part of a coordinated campaign to control the narrative inside the factory. Management doesn't like that IG Metall disputes Thierig's 'slacker theory' about high absenteeism.

Despite fierce resistance, the union manages to conduct a workplace survey. The council majority mocks it as 'just more flyer-nonsense.' Still, 1,200 employees take part, according to IG Metall. That's only 10 percent of the total workforce – but far more than expected. The results are damning. Tesla is seen as a workplace people don't stick with for long. Eighty-three percent say they often or very often feel overworked. Fifty-nine percent report regular physical pain from the job. Under the current conditions, only one in ten think they can make it to retirement.

Ruined Credit Rating

In Summer 2024, an email arrives reeking of trouble. It begins innocuously enough: 'We are a medium-sized taxi company, and I personally read *Handelsblatt* regularly,' writes an entrepreneur by the name of Dirk Holl. His next sentence grabs our attention. 'Last week, I intended to purchase a Tesla Model Y and finance it through AKF Bank, since Tesla does not have its own automotive bank. Normally, this wouldn't be an issue, but this time AKF Bank is refusing to disburse the loan amount to Tesla Germany GmbH, as Tesla has a credit rating of 600 (see attachment) according to Creditreform.'

Say what? In Germany's Bonitätsindex – Creditreform's credit-worthiness scale – a score of 600 is a catastrophe. It labels a company as high-risk, with an elevated probability of default. And yet Tesla – raking in billions – somehow bears that black mark?

We call Creditreform. A recent Berlin court case provides the explanation: Tesla Germany lost a dispute over an unpaid invoice and was ordered to provide a Vermögensauskunft – a sworn statement of assets – so creditors could gauge what might be seized. If a debtor fails to appear or refuses without excuse, the court can issue an arrest warrant and record the name in the debtors' register.

Which invoice went unpaid, and for how much, remains a mystery. When we reach out to Tesla, the company declines to comment. Did they dispute, ignore, or misplace the summons? We don't know – and Tesla has never replied to our inquiries.

It's baffling. Even in a lean 2024, Tesla's profits should cover any routine supplier bill. Instead, they risk being branded an unreliable partner, triggering automated credit-control measures that no corporate treasurer wants.

We run the story on a Friday, and at last a Tesla representative reaches out – uncharacteristically prompt. The matter, he says, was 'just a small unpaid invoice,' and it's now resolved. He promises to serve as our dedicated point of contact going forward.

We add his assurance to the article and the official credit rating does indeed improve. At the time of writing, Tesla Germany has received an excellent assessment from Creditreform, which estimates the probability that Musk's company will pay its bills at 99.42 percent. As for our supposed point man, that initial outreach gives way to radio silence. No matter how many questions we send, we never hear back from him.

He who does not work . . .

Around the same time we hear about Tesla's unpaid bill, something else catches our eye on Google Maps. Satellite images show rows upon rows of Teslas parked on a remote field in eastern Germany. Inside the company, we hear rumors: Grünheide's Gigafactory is planning a five-day production halt. Then, on June 8, Tesla announces something unexpected – a 13 percent price cut for the Model Y, once a bestseller.

We wonder if these dots are connected.

A bit of online sleuthing leads us to information about an event taking place right beside the mysterious Tesla parking site. The announcement reads: 'On June 8 and 9, 2024, Neuhardenberg Airport will celebrate its 90th anniversary. A diverse program attracts visitors from near and far.'

We decide to attend. Lars, a colleague based in Berlin, cancels his family plans for the morning and sets off to Neuhardenberg. What begins as a company investigation turns into a crash course in German history.

The Neuhardenberg airfield was built in 1934 as a secret military base for the German Luftwaffe. After Hitler's defeat, it became part of the National People's Army in East Germany, home to elite fighter and transport wings – including one responsible for flying the East German leadership. Today, the hangars are overgrown with grass. The contrast with what's parked outside could hardly be starker.

On the bus to the site, Lars films what he sees: an endless sea of Teslas, lined up six across in countless rows, shimmering under the unforgiving sun. Even during the festival, trucks continue to arrive, unloading more vehicles. A visitor quips, 'Looks like the airport is charging for parking.'

It's a surreal way to start the weekend. Beyond the forest of idle electric cars, there's a unicorn-themed bouncy castle, hot dog stands, and – for the anniversary – a Soviet-era

MiG-21 rolled out of its hangar like a Cold War ghost. At a small booth, two Tesla representatives offer test drives.

We spend the weekend piecing together a story that feels both absurd and emblematic: Tesla, the global tech giant, has produced more vehicles than it can sell. The field in Neuhardenberg is more than an overflow lot – it's a symbol of lost momentum.

At the Gigafactory in Grünheide, the mood darkens in the months that follow. Management continues to criticize the workforce but offers no clear plan for recovery. There's no new model in sight, and little indication of strategic direction from the top. While Elon Musk dominates headlines for his politics, Tesla's German workforce is still waiting for a roadmap – or even basic guidance – on the vehicles they build.

A sense of injustice begins to seep through the factory walls.

In the U.S., Tesla's board of directors grants Musk the largest bonus package ever awarded to a CEO. In Grünheide, employees receive a modest holiday gift: a pair of red socks bearing the white Giga logo.

One man got billions. Twelve thousand got socks.

As the year turns, things get worse. In previous months, we have reported on how factory chief André Thierig called his workers lazy, accused them of stealing his coffee mugs, and showed up at their homes when they called in sick. Now, Germany's Tesla boss seems to have a new message for his

12,000 employees – straight from the Bible, namely Paul the Apostle: 'He who does not work, neither shall he eat.'

'Doubts about submitted sick notes,' reads the subject line of numerous letters passed on to us in March. Officially, Tesla has resolved the issue of high absenteeism. As Thierig puts it in a staff meeting, he and his team 'invested very, very heavily in this topic and held "many, many conversations."' All this good work, he says, has led to a nearly 50 percent reduction in missed days. His conclusion: 'The work ethic, as far as I'm concerned, is definitely much improved.'

Now we hear the other side of the story. Employees who call in sick can expect a letter from Tesla – one that openly accuses them of faking their illness. 'Tesla has serious doubts about the actual existence of your incapacity for work,' one states. According to the company, there are 'well-founded doubts' that the employee is entitled to continued pay during their illness.

As we read these letters, disbelief sets in. In black and white, Tesla accuses its employees of inventing their illnesses. But it doesn't stop there. Management demands that workers release their treating physicians from doctor-patient confidentiality. Employees are expected to disclose 'which illness (with medical diagnosis) was the cause of their incapacity for work.' Until they comply: no further pay.

The ultimatum even catches IG Metall – battle-hardened as it is – off-guard. Regional director Dirk Schulze condemns the suspension of wages during illness as 'a completely unacceptable form of intimidation.' For employees and their families, he says, 'it's an incredible burden not to know whether they'll have enough income next month to pay their rent.' Tesla's methods, he says, are 'not only highly unprofessional and inhumane, but also counterproductive.'

As always, we ask Tesla for comment. As always, the company doesn't respond. Still, the story explodes – it's the most-read article on the *Handelsblatt* website by a mile. A day later, Thierig seems compelled to react. He grants an interview to another paper, defending the suspension of pay for sick employees as 'completely legitimate.'

In short, Thierig says he doesn't see what all the fuss is about: 'We're talking about around a dozen cases per month, so about 0.1 percent of the workforce where we choose to take this approach.' In such cases, he says, Tesla is simply 'questioning whether everything is as it seems.' It's then up to the employee to prove that 'the conditions for continued wage payment are met.'

Thierig does not explain why he refuses to accept official medical certificates and instead wants to hold one-to-one conversations with doctors. His tone suggests he's simply curious – nothing more.

The same can't be said of works council chair Michaela Schmitz. While her own boss openly admits to doubting his employees' sick notes, the issue seems to have passed her by. She, too, gives an interview after our publication and paints a sunny picture: 'I believe the working conditions in the Gigafactory are good,' she says, telling a local paper she has no information about the suspension of wages in case of sickness.

Schmitz, apparently, has no intention of finding out either. In April 2025, union-affiliated members of the works council submit a motion asking Tesla management to disclose exactly how many employees on sick leave are currently not being paid, and how that number has changed in recent months. Schmitz does not support the motion. Neither do any of the other non-unionized members.

Meanwhile, Thierig turns the next screw. 'We have determined that, according to our records, you received an overpayment. The reason for this is that you were mistakenly paid during your unpaid leave,' reads another letter, shared by distressed employees. In Tesla's record, they are not only unpaid – they now owe money.

According to IG Metall, Tesla's management has made private offers to affected employees: those who immediately sign a termination agreement will have their supposed debts waived. But it doesn't always go that smoothly.

The union is gaining ground at the German Gigafactory. Workers seem less inclined to believe Musk's tale of IG Metall being a spy operation for the old auto industry – and more willing to fight for labor standards which have a long tradition in Germany. Thousands have signed a petition demanding just that. But management's current behavior is only fueling frustration – and resistance.

So, they follow the union's advice: never sign a termination agreement without a prior legal review. 'These so-called overpayments are, in almost every case, completely baseless accusations,' says regional director Dirk Schulze, emphasizing that IG Metall is offering legal support. According to him, the union has already recovered nearly half a million euros from Tesla: 'Money the workers earned — money that would otherwise have ended up in the pocket of the richest man in the world.'

In 2025, the rift between Musk and his workforce is widening fast. Not long ago, he stood in front of his employees in Grünheide, speaking with enthusiasm about his vision for Tesla: a place not just to work, but to have fun. But if Musk once promised that exceptional performance would lead to exceptional rewards, the reality is more prosaic – the employees are left holding a pair of socks while he counts the billions granted to him by the board.

Tesla's history in Germany spans just three years. But the early excitement has faded. Management is locked in conflict with the union. Workers are dismissed as slackers, and trust is eroding fast. Many of the employees we speak with now see only two options: accept this harsh new reality – or walk away.

Musk once described his factories as hell.

In Grünheide, at least, there's an exit.

CHAPTER 9

Life and Death on Autopilot

Of all the conversations we've had while reporting on the Tesla Files, the one with Rita Meier* is among the hardest. It's a Monday afternoon in June 2023 when the then 45-year-old joins us for a video call. She tells us about the last time she said goodbye to her husband, Stefan,* five years earlier. He was leaving their home near Lake Constance, Germany, heading for a trade fair in Milan, Italy. Afterward, he planned to check out the boat he had just purchased, docked somewhere on the Ligurian coast.

Meier recalls how he hesitated between taking his Model S or her BMW. He had never driven the Tesla that far before. He checked the route for charging stations along the way to Italy, and ultimately decided to try the

---

* Name changes marked with an asterisk throughout.

Tesla. Rita had a bad feeling. She stayed home with their three children, the youngest not even a year old.

At 3:18 p.m. on May 10, 2018, Stefan Meier loses control of his Model S on the A2 highway near the Monte Ceneri tunnel. Traveling at around 100 km/h, he plows through several warning markers and traffic signs before crashing into a slanted guardrail. 'The collision with the guardrail launches the vehicle into the air, where it flips several times before landing,' investigators will write later. The car comes to rest more than 70 meters away, on the opposite side of the road, leaving a trail of wreckage. According to eyewitnesses, the Model S bursts into flames while still airborne. Several passersby try to open the doors and rescue the driver, but they can't unlock the car. When they hear explosions and see flames through the windows, they retreat. Even the firefighters, who arrive about 20 minutes later, can do nothing but watch the Tesla burn.

At that moment, Rita Meier is unaware. She tries calling her husband, but he doesn't pick up. That alone doesn't worry her. What raises the alarm is that he still hasn't returned her call hours later – highly unusual for the devoted father of three. She attempts to track his car using Tesla's app. It no longer works. By the time police officers ring her doorbell late that night, Meier is already bracing for the worst.

The crash makes the headlines the next morning as one of the first fatal Tesla accidents in Europe. Tesla releases a statement to the press saying the company is 'deeply saddened' by the incident. 'We are working to gather all the facts in this case and are fully cooperating with local authorities.'

To this day, Meier still doesn't know why her husband died. She keeps everything the police gave her after their inconclusive investigation. The charred wreck of the Model S sits in a garage Meier rents specifically for that purpose. The scorched phone, which she had forensically analyzed at her own expense with little result, rests in a drawer at home. Maybe someday all this will be needed again,' she says. Meier hasn't given up hope of uncovering the truth. She isn't seeking publicity or compensation, she says. She simply wants to understand why her husband died.

Meier is one of the many people who reached out to us after our first 'Tesla Files' report on problems with the Autopilot system. Like hundreds of others, she entered the vehicle identification number of her husband's Model S into the response form we published on the *Handelsblatt* website. That's when she discovered that the Tesla Files contained data related to the car. In her first email, she wrote: 'You can probably imagine what it felt like to read that.'

There isn't much information. Just an Excel spreadsheet titled 'Incident Review.' A Tesla employee noted that the odometer on Stefan Meier's car stood at 4,765 miles at the time of the crash. The entry was catalogued just one day after the fatal accident. In the comment field: 'Vehicle involved in an accident.' The cause of the crash remains unresolved to this day. And yet, in Tesla's internal system, a company employee marked the case as 'resolved.'

We can't help but wonder how that word must sit with his widow. For five years, Rita Meier has been searching for answers. For five years, she's been trying to make sense of why the father of her children is no longer at the dinner table. After Stefan's death, she took over the family business – a timber company with 200 employees based in Tettnang, Baden-Württemberg. She had to move on. Questions be damned.

As journalists, we're used to tough interviews. But this one is different. We have to strike a careful balance – between empathy, which we're trained to switch off when necessary, and the tough, persistent questioning that good reporting demands. 'Why are you convinced the Tesla was responsible for your husband's death?' we ask. 'Isn't it possible he was distracted – maybe looking at his phone?'

We're struck by the calm, collected way Meier answers us. Her account is structured and precise. Only once does the toll become visible – when she describes how her husband's body burned in full view of the firefighters. Her eyes fill with tears, and her voice cracks. She apologizes briefly, turning away from the camera. When she collects herself, she says she has nothing left to gain – but also nothing to lose. That's why she reached out to us.

We're unsure how to respond. We promise to look into the case.

Meier's story strikes a chord. All of us on Team Tesla have families. Meier's tragedy reminds us how fragile that happiness can be. All the abstract numbers we've been dealing with over the past months – the countless Excel spreadsheets filled with complaints and crash reports: Some of them represent horrible tragedies.

Most Tesla customers walk away from their accidents. Some don't. And while an anonymous employee quietly closes a fatal crash case in the system after just a few days, everything changes for those left behind. Partners and children, parents and grandparents, friends and loved ones – their lives are forever marked by an unbearable void.

Fatal Silence

Rita Meier isn't the only widow to approach us. Our investigation into flaws in Tesla's Autopilot has made waves. Disappointed customers, current and former employees, analysts, and lawyers are sharing links to our reporting. Many write to us. Often, we get the sense that they see *Handelsblatt* as a kind of David taking on Goliath. More than once, someone writes that it's about time someone stood up to Tesla – and to Elon Musk.

Meier, too, shares our articles and the callout form with

others in her network – including people who, like her, lost loved ones in Tesla crashes and have been searching for allies ever since.

One of them is Anke Schuster.* Like Meier, she lost her husband in a Tesla crash that defies explanation. Schuster has spent years chasing answers. And like Meier, she finds her husband's Model X listed in the Tesla Files. Once again, the incident is marked as 'resolved' – with no indication of what that actually means.

'In the black Tesla Model X with the VIN listed above, my husband died in an unexplained and inexplicable accident,' Schuster writes in her first email. Her dealings with police, prosecutors, and insurance companies, she says, have been 'hell.' No one seems to understand how a Tesla works. 'I lost my husband. His four daughters lost their father. And no one ever cared.'

Her husband Oliver was a tech enthusiast, fascinated by Elon Musk. A hotelier by trade, he owned no fewer than four Teslas. He loved the cars. She hated them – especially the Autopilot. She always found it suspicious. The way the software seemed to make decisions on its own never sat right with Schuster. Today, she sees her instincts confirmed in the worst way.

Oliver was returning from a business meeting on April 13, 2021, when his Model X veered off the highway B194 between Loitz and Schönbeck in northeastern Germany. It was 12:50 p.m. when the car left the road and crashed into a tree. Schuster started to worry when Oliver missed a scheduled bank appointment. She tried to track the vehicle but found no way to locate it. Even calling Tesla led nowhere. That evening, the police told her the brutal truth: her husband's car had burst into flames after the crash. He had burned to death – with the fire brigade watching helplessly.

The crashes that killed Meier's and Schuster's husbands are more than three years apart. The parallels are chilling. We examine accident reports, eyewitness accounts, crash-site photos, and correspondence with Tesla. In both cases, investigators requested vehicle data from Tesla. The company didn't provide it. In Meier's case, Tesla staff claimed no data was available. In Schuster's, it said there was no relevant data.

For us, these emails from the widows mark the beginning of one of the most emotionally intense Tesla investigations we've ever undertaken. Over the next two years, we'll speak with crash victims, grieving families, and experts around the world. What we uncover is an ominous black box – a system not only built to collect and control every byte of customer data, but designed to safeguard Musk's vision of autonomous driving. Critical information is sealed off from public scrutiny.

Let's start at the beginning. For a company that markets its cars as computers on wheels, Tesla's claim that it had no data available is surprising. Musk has long described Tesla vehicles as part of a collective neural network – machines that continuously learn from one another. Think of the iconic Borg aliens from the science fiction series *Star Trek*. Musk envisions his cars, like the Borg, as a collective – operating as a hive mind, each vehicle linked to a unified consciousness.

When a journalist asks him in October 2015 what makes Tesla's driver-assistance system different, he replies: 'The whole Tesla fleet operates as a network. When one car learns something, they all learn it. That is beyond what other car companies are doing.' Every Tesla driver, he explains, becomes a kind of 'expert trainer for how the autopilot should work.'

Vast amounts of video footage from millions of Teslas worldwide are intended to make the system smarter with every mile. In 2019, Musk unveils a project called 'Dojo' – a supercomputer designed to process that flood of data at unprecedented speed. He calls it a 'beast,' promising it would represent a 'quantum leap' for self-driving technology.

In June 2023, Elon Musk doubles down: 'With respect to Autopilot and Dojo, in order to build autonomy, we obviously need to train our neural net with data from millions of vehicles. This has been proven over and over again – the more training data you have, the better the results.'

He emphasizes that no other company has more data-collecting vehicles on the road than Tesla. 'There's just no substitute for a massive amount of data. And obviously, Tesla has more vehicles on the road collecting this data than all of the other companies combined,' Musk claims. 'I really just don't know how anyone could do what we are doing – even if they had our software and our computer – if they did not have the training data.'

On its website, Tesla boasts that its AI 'provides a view of the world that a driver alone cannot access, seeing in every direction simultaneously, and on wavelengths that go far beyond the human senses.' The company claims that this forms the foundation of Autopilot's learning capabilities. Accidents may still happen – but only for a while longer. Because every Tesla is connected, the company has access to data from millions of vehicles. It uses this network to 'understand the different ways accidents happen. We then develop features that can help Tesla drivers mitigate or avoid accidents.'

According to Elon Musk, the eight cameras in every Tesla transmit more than 160 billion video frames per day to the company's servers. In its owner's manual, Tesla states that its cars may collect even more: 'analytics, road segment, diagnostic, and vehicle usage data,' all sent to headquarters to improve product quality and features like Autopilot. The company claims it learns 'from the experience of billions of miles that Tesla vehicles have driven.'

It is a powerful promise: a fleet of millions of cars, constantly feeding raw information into a gargantuan processing center. Billions – trillions – of data points, all in service of one goal: making cars drive better, and keeping drivers safe. As the year turned, Musk got a chance to show the world what he meant.

On January 1, 2025, at 8:39 a.m., a Cybertruck explodes outside the Trump Hotel in Las Vegas. The man behind the incident – U.S. special forces veteran Matthew Livelsberger – had rented the vehicle, packed it with fireworks, gas canisters, and grenades, and parked it in front of the building. Just before the explosion, he shot himself in the head with a .50 caliber Desert Eagle.

'This was not a terrorist attack, it was a wakeup call. Americans only pay attention to spectacles and violence,' Livelsberger wrote in a letter later found by authorities. 'What better way to get my point across than a stunt with fireworks and explosives.'

The soldier miscalculated. Seven bystanders suffered minor injuries. The Cybertruck was destroyed, but not even the windows of the Trump Hotel shattered. Instead, with his final act, Livelsberger revealed something else entirely: just how far the arm of Tesla's data machinery can reach. 'The whole Tesla senior team is investigating this matter right now,' Musk writes on X just hours after the blast. 'Will post more information as soon as we learn anything. We've never seen anything like this.'

Later that day, Musk posts again. Tesla has already analyzed all relevant data – and is ready to offer conclusions.

'We have now confirmed that the explosion was caused by very large fireworks and/or a bomb carried in the bed of the rented Cybertruck and is unrelated to the vehicle itself,' he writes. 'All vehicle telemetry was positive at the time of the explosion.'

Suddenly, Musk isn't just a CEO. He is an investigator. He instructs Tesla technicians to remotely unlock the scorched vehicle. He hands over internal footage captured up to the moment of detonation. Musk's fervor makes the police look like bystanders. In the end, when handed everything there was to know about the attack, the officer in charge proves to be well mannered.

'I have to thank Elon Musk, specifically,' says Las Vegas Metropolitan Police Department Sheriff Kevin McMahill to reporters. 'He gave us quite a bit of additional information.'

In that moment, what was meant as an act of destruction transforms into something different: A demonstration of Musk's power. The Tesla CEO has turned a suicide attack into a showcase of his superior technology.

And yet, there are critics even in the moment of glory. 'It reveals the kind of sweeping surveillance going on,' warns David Choffnes, executive director of the Cybersecurity and Privacy Institute at Northeastern University in Boston when reached by a reporter. 'When something bad happens, it's helpful, but it's a double-edged sword. Companies that collect this data can abuse it.'

There are other examples of what Tesla's data collection makes possible. We find the case of David and Sheila Brown, who died in August 2020 when their Model 3 ran a red light at 114 mph in Saratoga, California. Investigators managed to reconstruct every detail thanks to Tesla's vehicle data. It shows exactly when the Browns opened a door, unfastened a seatbelt, and how hard the driver pressed the accelerator – down to the millisecond, right up to the moment of impact. Over time, we find more cases, more of such detailed accident reports. The data is definitely there – until it isn't.

Tesla promises customers transparency and full control over the data their cars collect. Musk says so himself: 'Tesla releases critical crash data affecting public safety immediately & always will.' In a tweet from April 2, 2018, he writes: 'To do otherwise would be unsafe.'

So, we've seen what true transparency looks like. But what happened in the cases of Rita Meier and Anke Schuster? Why are these widows still waiting for answers about their husbands' deaths? Presumably, their cars transmitted data – just not at the moment of impact. How does that square with Musk's promise of a constantly evolving fleet of intelligent machines? In a company built to capture everything, why is there so little insight to be gained when two Tesla customers die using its products?

Hacking Elon's Code

We keep chasing that question. By now, we know a lot about how Musk's cars work – but the way they handle data still feels like a black box. How is that data stored? At what moment does the onboard computer send it to Tesla's servers? We need to understand this better. So we set out to talk to experts – people who are demonstrably independent.

We find them at the Technical University of Berlin. Three PhD candidates – Christian Werling, Niclas Kühnapfel, and Hans-Niklas Jacob – made headlines for hacking Tesla's Autopilot hardware. Their tools? Just €600 worth of gear. Their method? A brief voltage drop on a circuit board – just enough to trick the system into opening up.

The security researchers uncovered what they call 'Elon Mode' – a hidden setting in which the car drives fully autonomously, without requiring the driver to keep his hands on the wheel. They also managed to recover deleted data, including video footage recorded by a Tesla driver. And they traced exactly what data Tesla sends to its servers – and what it doesn't.

Werling, Kühnapfel, and Jacob know their stuff. At a tech conference, they present the core findings of their research – and explain what had driven them: despite all the supposed sophistication of the software, crashes keep happening in Autopilot mode. And Tesla, they say, withholds data that might help explain why. That gap between promise and reality is what pushed them to dig deeper.

We believe their insights can help us, so we arrange a meeting. We tell them about our investigation: the fatal crashes, the missing data, the widows we've spoken to. We share what we've seen in the Tesla Files – how cases are marked 'resolved' even when no answers have been given.

The doctoral candidates' explanations are highly technical. We interrupt often – asking them to clarify, slow down, repeat. It's hard to follow – but by the end of the meeting, one thing is clear: we're on to something important.

The hackers explain that Tesla stores data in three places. First, on a memory card inside the onboard computer – essentially a running log of the vehicle's digital brain. Second, on the Event Data Recorder – a 'black box' that captures a few seconds before and after a crash. And third, on Tesla's servers, assuming the vehicle uploads them.

The researchers tell us they've found an internal database embedded in the system – one built around so-called 'trigger events.' These events determine what data the car stores, when it stores it, and where it's sent. If, for example, the airbag deploys or the car hits an obstacle, the system is designed to save a defined set of data to the black box – and transmit it to Tesla's servers.

Unless the vehicles were in a complete network dead zone, one thing seems clear: in both the Meier and Schuster cases, the cars should have recorded and transmitted that data.

No Access, No Answers

Now that we have a basic grasp of how Tesla's data systems function, we want to understand who inside the company actually works with that data. We examine testimony from Tesla employees in court cases related to fatal crashes. They describe how their departments operate. We cross-reference their statements with entries in the Tesla Files.

A pattern takes shape: one team screens all crashes at a high level, forwarding them to specialists – some focused on Autopilot, others on vehicle dynamics or road grip. We'd already noticed, while reporting on phantom braking and sudden acceleration, that the same staff names keep surfacing. Now we see there's also a group that steps in whenever authorities request crash data.

We compile a list of employees relevant to our reporting. Some we try to reach by email or phone. For others, we show up at their homes. If they're not there, we leave handwritten notes. No one wants to talk.

We also know the names of Tesla's lawyers who told authorities in the Meier and Schuster cases that no data was available. We don't contact them – not yet. Years of journalism have taught us: If we want someone to open up, we'll need more information first.

So, we search for other crashes. Fortunately, Tesla critics have been tracking fatal accidents online for years. We go through their records and select a few cases. Then we reach out to the investigating authorities. We're also watching for new incidents. Every morning, we scan news alerts and police reports for crashes involving Teslas – cars that veered off the road, collided with trees, or slammed into walls. Our questions are always the same: Was the cause of the crash determined? And if data was requested from Tesla – did the company deliver?

It doesn't take long to find more cases. One involves Hans von Ohain, a 33-year-old Tesla employee from Evergreen, Colorado. On May 16, 2022, he crashes into a tree on his way home from a golf outing. The car bursts into flames. Von Ohain dies at the scene. His passenger survives.

Von Ohain had been drinking. His passenger tells police that he activated Full Self-Driving – a feature designed to enable fully autonomous driving. Tesla, however, says it can't confirm whether the system was engaged – because no vehicle data was transmitted for the incident.

Then, in February 2024, Elon Musk himself steps in. The Tesla CEO claims von Ohain had never downloaded the latest version of the software – so it couldn't have caused the crash. Friends of von Ohain, however, tell U.S.

media that he had shown them the system. His passenger that day, who barely escaped the wreck with his life, told reporters that hours earlier the car had already driven erratically by itself. 'The first time it happened, I was like, "Is that normal?" he recalled asking von Ohain. 'And he was like, "Yeah, that happens every now and then."'

His account was bolstered by von Ohain's widow, who explained to the media how overjoyed her husband was to be working for Tesla. Reportedly, von Ohain received the Full Self-Driving system as a perk. His widow recalls how he would use the system almost every time he got behind the wheel: 'It was jerky, but we were like, that comes with the territory of new technology. We knew the technology had to learn, and we were willing to be part of that.'

The Colorado State Patrol investigates but closes the case without blaming Tesla. It reports that no usable data was recovered.

A similar pattern emerges in the crash near Dobbrikow, Germany, in August 2022, when two 18-year-olds burn to death inside a Model S before rescuers can intervene. An investigator from Dekra, one of Germany's leading testing authorities, later writes to prosecutors that any relevant data 'would have to come from the manufacturer.' But, he adds, 'experience shows that vehicle manufacturers are often unwilling to comply with such requests.' The best chance, he notes, is when authorities specify precisely which data they want.

The prosecutor follows that advice. Tesla responds with a formal letter, stating it is 'ready and willing' to support the investigation 'as best as possible.' And yet: there is 'no data for the period in question' on the company's servers.

In many crashes when Teslas inexplicably veer off the road or slam into stationary objects, investigators don't even request data from the company. When we ask authorities why, there's often silence. Our impression is that many prosecutors and police officers aren't even aware that asking is an option. In other cases, they act only when pushed by victims' families. That realization is one of the more sobering outcomes of this investigation: if you want answers, you can't count on others to find them. Not even the authorities whose job is exactly that. Some families say investigators refused to request data – and instead asked whether the crash might have been a suicide.

In the Meier case, Tesla tells authorities in a letter dated June 25, 2018, that the last complete set of vehicle data was transmitted nearly two weeks before the crash. The only data from the day of the accident is a 'limited snapshot of vehicle parameters' – taken 'approximately 50 minutes before the incident.' However, this snapshot 'doesn't show anything in relation to the incident.' As for the black box, Tesla warns that the storage modules were likely destroyed, given the condition of the burned-out vehicle. Data transmission after a crash is possible, the company says – but in this case, it didn't happen. In the end, investigators can't even determine whether driver-assist systems were active at the time of the crash.

The Schuster case plays out similarly. Prosecutors in Stralsund are baffled. The road where the crash happened is straight. The asphalt was dry, the weather at the time of the accident was clear. Schuster keeps urging the authorities to examine Tesla's telemetry data. Eventually, a police officer – clearly frustrated – writes to the crash investigator: 'Do you know anything about Tesla vehicles being monitored electronically at all times? [. . .] The widow wants answers as to why the car drove into the tree without braking. I need to check whether the DA wants to pursue that line of questioning.'

The crash investigator replies that it is 'entirely possible (and likely)' that Tesla has 'a large amount of data related to this incident.' The problem, he adds, is access. It's unclear whether a German court order would even be enforceable in the U.S.

Prosecutors don't rule out a vehicle error. They formally request all data recorded by Schuster's car on the day of the crash. The chief prosecutor writes that the data is 'necessary to determine whether a technical defect can be ruled out.' It takes Tesla more than two weeks to respond – and when it does, the answer is both brief and bold. The company doesn't say there is no data. It says: 'no relevant data.'

The authorities' reaction leaves us stunned. We expect prosecutors to push back – to tell Tesla that deciding what's relevant is their job, not the company's. But they don't. Instead of insisting, they close the case. No data means no evidence of fault. That's the moment we realize just how impenetrable Tesla's data systems really are. Anyone requesting vehicle data – from customers to families to public prosecutors – has to take the company at its word. There's no way to verify what's withheld.

It's a sobering realization. Especially because we've seen Tesla handle its data unfairly elsewhere.

Among our readers are many customers currently battling Tesla in court. Some contact us hoping for ammunition to use in their lawsuits. We always decline. We're journalists, not partisans. But we try to talk to everyone who might have a story. Our thinking: every frustrated customer might be sitting on something important.

That's how we come across the case of Emil Dupont,* an IT consultant from Belgium. He buys a Model S in mid-2018 – and complains of defects from day one. Seven repair attempts later, the issues persist. Tesla offers to take the car back for the full purchase price, but never follows through. The case ends up in court.

Dupont wins. The appeals court in Antwerp orders Tesla to refund him €158,600. The ruling is a bombshell – the highest-profile court decision in Europe against Tesla over a consumer complaint. The court's reasoning reads like a direct rebuke of Elon Musk and his Autopilot hype.

The Belgian judge attests that Dupont's Tesla suffers from 'serious defects in driving comfort and safety.' The IT consultant's recurring issues, the court finds, demonstrate that 'the vehicle is not suited for its intended normal use.'

No Relevant Data

The Belgian court case is more than just another lawsuit. It reveals how Tesla tried to block access to critical data. The court appointed an expert to inspect the vehicle. He informed the company which data he needed – and why it was essential to his investigation.

At first, Tesla claimed the log files couldn't be extracted directly from the vehicle. Then the company shifted its stance, admitting the data was stored in the car – but it declined to retrieve it. Later, Tesla changed its explanation again, saying the data was stored only temporarily before being uploaded to the cloud. At a subsequent technical meeting, Tesla engineers offered yet another version: the data, they said, remained in the vehicle for 18 months before being overwritten.

At that meeting, Tesla refused to extract and share the log data – despite the expert stressing that accessing all of it was necessary to complete his work. The expert later noted that Tesla had unilaterally accessed the car's trip logs without informing either the parties or the court. In his report, he wrote that this conduct had 'hindered the smooth progress of the investigation.'

It was only 'after considerable pressure and heated debate' between the parties, their lawyers, and the expert that Tesla released any data at all – but even then only 'a few partial, very limited, incomplete, and selectively filtered logs.' Log data from the period prior to the initial expert meeting and test drive was not provided, nor was other specific information requested by the expert shared or fully answered. As a result, the expert was unable to meaningfully verify whether the defects and shortcomings cited by the appellants were reflected in the vehicle's log data.

After all this hassle, the court rejected Tesla's claim that it had 'fully cooperated with the investigation,' noting that this contradicted with 'the available documentation, the evidence on file, and the content of the expert's report.'

The hackers from TU Berlin point us to a study by the Netherlands Forensic Institute (NFI), an independent division of the Ministry of Justice and Security. In October 2021, the NFI published findings showing it had successfully accessed the onboard memories of all major Tesla models. The researchers compared their results with accident cases in which police had requested data from Tesla. Their conclusion: while Tesla formally complied with those requests, it omitted large volumes of data that might have proved useful.

Tesla's credibility also takes a hit in a report released by the U.S. National Highway Traffic Safety Administration (NHTSA) in April 2024. The agency concluded that Tesla failed to adequately monitor whether drivers remain alert and ready to intervene while using its driver-assist systems. It reviewed 956 crashes, field data, and customer communications.

The NHTSA pointed to 'gaps in Tesla's telematic data' that made it impossible to determine how often Autopilot was active during crashes. If a vehicle's antenna was damaged or it crashed in an area without network coverage, even serious accidents sometimes went unreported. Tesla's internal statistics include only those crashes in which an airbag or other pyrotechnic system deployed – something that occurs in just 18 percent of police-reported cases. This means the actual accident rate is significantly higher than Tesla discloses to customers and investors.

There's more. Two years prior, the NHTSA had flagged something strange – something suspicious. In a separate report, it documented 16 cases in which Tesla vehicles crashed into stationary emergency vehicles. In each, Autopilot disengaged 'less than one second before impact' – far too little time for the driver to react. Critics warn that this behavior could allow Tesla to argue in court that Autopilot was not active at the moment of impact, potentially dodging responsibility.

The YouTuber Mark Rober, a former engineer at the U.S. space agency NASA, replicates this behavior in an experiment on March 15, 2025. He simulates a range of hazardous situations, in which the Model Y performs significantly worse than a competing vehicle. The Tesla repeatedly runs over a crash-test dummy without braking. The video goes viral, amassing more than 14 million views within just a few days.

The real surprise comes after the experiment. Fred Lambert, who writes for the blog *Electrek*, points out that Autopilot appears to disengage less than a second before the staged collision – just as the NHTSA previously documented. 'Autopilot appears to automatically disengage a fraction of a second before the impact as the crash becomes inevitable,' Lambert notes, referencing the footage.

And so, the controversy continues. The NHTSA has been at odds with Tesla for years. One reason: the company's persistent reluctance to release data. That alone should raise red flags – not just for regulators, but for anyone trying to judge how safe these cars really are. It prompts a fundamental question: If the data support Tesla's claims about Autopilot performance, why keep it hidden?

Tesla's secrecy also shows up in the Tesla Files. We search for keywords like 'black box,' 'crash data,' and

'NHTSA' – and spend weeks combing through the results. Eventually, one name stands out: Steve Bradley, a UK-based engineer who keeps cropping up in internal discussions about crash analysis.

Bradley, we come to understand, is Tesla's fixer – the one they bring in when the stakes are highest. Internal memos and emails show that the company tasks him with building a team responsible for all liability-related cases across the EMEA region – Europe, the Middle East, and Africa. Whenever a crash in Europe involves sudden acceleration or braking, the case lands with his team. Soon we're able to match individual staff members to this group – and confirm that both the Meier and Schuster cases are handled by Bradley's team. In both instances, the result is the same: the case is marked 'resolved.'

Secrets and Subpoenas

We read every file that mentions Bradley's name and begin to piece together his role inside the company. The documents suggest he no longer works at Tesla and now lives in a small cottage in England. For a moment, we feel a flicker of hope: a man this central to crash investigations, now outside the company – he might be the key to understanding how Tesla handles sensitive incidents.

But Bradley won't speak to us.

Still, the Tesla Files offer a window into his world. They show what kinds of cases he handled, how he built his team, what internal processes he tried to improve – and where he ran into the limits of Musk's system.

One reason Bradley had so much on his plate is that he was in charge of the SII project. The letters stood for Safety Incident Investigation – a program meant to create a secure environment and a standardized process for handling safety-related incidents.

In internal memos, Bradley laid out his goals – and his frustrations. He was concerned that not all teams understood which events could trigger product liability. To address that, he pushed for proper training materials and clearer protocols.

He joined calls and meetings across departments to offer what he called 'guidance on how to handle cases of a technical nature' – often working with Tesla's legal team. He also consulted with colleagues in the U.S. to help develop a global framework for evaluating safety incidents and structuring the company's response.

One question persistently weighed on him: How does Tesla prevent critical information from leaking? There were 'some issues with smoothly granting access to a project and identifying appropriate watchers,' Bradley noted. Later, he reported: 'Security changes implemented. Now able to lock down each ticket to only watchers.' But internal secrecy had its downsides: 'The restricted nature of the portal causes the greatest confusion as it is not straight-forward to add people to tickets.'

On April 1, 2020, Bradley recorded that the project had become 'the primary portal for secure investigation of sensitive incidents that may entail concerns over product liability or integrity.' He noted that most safety incidents had been moved 'out of email into a secure area.' His team had created 56 tickets covering 'a wide range of concerns.' A core team was formed for 'top-level management of concerns,' and access restrictions were introduced to 'maintain confidentiality.'

At times, Bradley sounded exasperated. 'Not making much usable progress on this,' he wrote on June 25, 2020 in a memo about handling especially sensitive incidents involving Tesla vehicles. The reason: Tesla's U.S. leadership resisted full transparency in documentation. 'There is some concern on the U.S. side about having particular issues detailed in articles as this opens up the possibility of this information being requested under subpoena.'

Say what? We read those lines twice. Then again. This is it – proof that Tesla withholds data.

A subpoena allows a U.S. court to compel third parties to provide information relevant to a case. That includes forcing a company to produce internal documents, emails, or technical data during legal proceedings. Unlike a search warrant, authorities don't show up in person to seize files. Still, a subpoena is a binding legal order. Ignoring or defying one carries consequences – ranging from fines to jail time. A judge may treat noncompliance as grounds for a kind of default judgment – ruling against the party that refused to cooperate, no matter the merits of their case.

Bradley's memo focused on how Tesla service technicians could better identify incidents involving 'potential Product Liability claims' – and handle them more effectively. He wanted to review all internally recorded cases that 'have a high propensity to be subject to liability/ integrity claims.' His goal: to ensure that employees had 'clear and appropriate guidance' for managing such cases – from documentation to escalation. But serious gaps remained, he wrote – for example, in cases involving 'non-deployment of airbags etc.'

What clearly bothered Bradley most was the reluctance of Tesla's U.S. leadership. They were stalling progress, he believed, not out of technical necessity – but out of fear. Fear of lawsuits. Fear of subpoenas. 'This is frustrating,' he wrote, 'as I think there is more to be gained from having a clear direction for Service on how to direct these concerns. Looking to review this again and understand the concerns better – and how to overcome them.'

To us, these lines are confirmation. We now know that Tesla embellished its crash statistics – as the U.S. agency NHTSA demonstrated; that it obstructed a legal investigation – as the Belgian court concluded. And that it was preoccupied with the risk of sensitive data leaking.

Tesla worked actively to shield such information from outsiders. Among the data the company sought to keep out of subpoena range were, in all likelihood, pieces of potential evidence. Bradley's notes show that Tesla wasn't above stalling internal processes – if that's what it took to protect itself from liability.

We contacted Tesla multiple times with questions about the company's data practices. We asked about the Meier and Schuster cases – and what it means when fatal crashes are marked 'resolved' in Tesla's internal system. We asked the company to respond to criticism from the U.S. traffic authority, to the findings of Dutch forensic investigators, and to its failure to release vehicle data in the Belgian case involving Emil Dupont and his Model S. We also inquired why Tesla doesn't simply publish crash data, as Elon Musk once promised. And of course, we asked whether the company considers it appropriate to withhold information from potential U.S. court orders.

Not that we expected anything different – but, for the record: Tesla has not responded to any of our questions.

Musk's Black Box

At the end of this investigation, two things must be kept separate. The original cases of Meier and Schuster remain blind spots. That's frustrating for us – but we have to admit that we cannot give the widows the answers they've been hoping for all these years. After all our research, we simply cannot say whether or not there truly was no data from the crashes that killed their husbands, as Tesla claimed at the time. We don't even know if anyone exists who could say so with certainty. Rita Meier and Anke Schuster will likely never know why their husbands lost control of their Teslas. Why they had to die.

What's left? Elon Musk boasts about the vast amount of data his cars generate – data that, he claims, will not only improve Tesla's entire fleet, but also revolutionize road traffic. Onward, to a brighter future.

Well, if Musk wants to convince the world of that, he should prove it. He should show that the technology works. That Tesla's artificial intelligence truly surpasses human drivers. The data to demonstrate this exists. But, as we have witnessed again and again in the most critical of cases, Tesla refuses to share it. Does Musk have doubts?

The Tesla CEO frames his company as a force destined to make the world safer. In that vision, there is no room for doubt. That may be one reason Musk is willing to endure bad press and accept the harsh verdict in Belgium, rather than opening Tesla up. In the end, it seems, the company would rather be seen as secretive than risk admitting to flaws.

Musk is right about one thing: data is power. And power belongs to those who control it. Only they get to decide what it means. Musk has built a system that forces everyone to rely on his word – on his version of the data. Not just customers, but regulators. And widows like Rita Meier and Anke Schuster, still searching for answers. For themselves. For their children

Tesla's handling of crash data affects even those who never wanted anything to do with the company. Every road user trusts the car in front, behind, or beside them not to be a threat. Trusts its driver to be responsible. Does that trust still stand when the car is driving itself?

Tesla says yes, pointing to its crash statistics. Yet, according to the U.S. traffic authority NHTSA, those figures are flawed. Even in the accidents Tesla does report, the details remain hidden from the public.

Internally, we called our investigation into Tesla's crash data 'Black Box.' At first, because it dealt with the physical data units built into the vehicles – so-called black boxes. The term comes from aviation, where flight recorders are designed to help uncover the cause of a crash. But the devices Tesla installs hardly deserve the name. Unlike those in planes, they're not fireproof – and in many of the cases we examined, they proved useless.

Over time, we came to see that 'Black Box' held a second meaning. A black box, in common parlance, is something closed to the outside. Something opaque. Unknowable. And while we've gained some insight into

Tesla as a company, its handling of crash data remains just that: a black box.

Only Tesla knows how Elon Musk's vehicles truly work. And yet, today, more than 7 million of them share our roads.

CHAPTER 10

Leviathan

Barely slept last night. Wednesday, November 6, 2024: The morning after the U.S. election. We're in an office on the fifth floor of the *Handelsblatt* building. Push notifications keep lighting up our phones. Then comes the alert: the key swing state of Pennsylvania is called for Donald Trump. Moments later, the Republican declares himself the winner.

We discuss Elon Musk's America political action committee (PAC), which had declared Trump president just two hours earlier. We talk about the photos from that night, showing Trump at his Mar-a-Lago residence in Florida, celebrating with Musk by his side. How surreal it all feels. How dystopian.

'We have a new star. A star is born, Elon!' says Trump in his victory speech at the Palm Beach Convention Center, not far from his villa. The crowd cheers. 'He's an amazing guy. We were sitting here tonight. He saved a lot of lives.' Trump praises Musk's rockets and satellites. 'He's a character, he is a super-genius. We have to protect our geniuses. We don't have that many of them.'

Musk made it.

While others vote for a president, Musk bought one. He poured almost $300 million into Trump's campaign. In 2024, Musk turned his messaging platform X into a propaganda machine, with himself as its loudest voice. In the final weeks, he toured Pennsylvania for Trump. No entrepreneur has ever backed a political candidate quite like this.

Now his investment is paying off.

Even before Trump's speech ends, trading opens in Frankfurt. Tesla shares begin climbing on our monitors. Later that day, Musk meets with the president-elect to discuss the cabinet, a first to-do list, and his own role: Chief Cost Cutter. Trump wants to give Musk what he came for. What he gambled on. A central post in the new administration. A position with leverage over matters close to him. Every billionaire's dream.

'Never bet against Elon,' PayPal co-founder Peter Thiel once said. These days, he must feel validated.

The world, on the other hand, must wonder how Musk will wield his new instrument of power. His business interests stretch far – from federal contracts to regulatory influence, labor law, environmental rules. Musk's reach has no borders. He won't hold office. Won't be tied to any domain. He'll operate behind the scenes – where ethics codes can't constrain him. Not even when he moves to shut down investigations into himself or his companies.

There's no doubt he'll try. No CEO is pushing his own agenda so brazenly. His mother calls it self-defense. 'Well, he had no choice,' Maye Musk tells Fox News the night before the election. She claims Kamala Harris wanted to shut down X. That's false, but she continues: 'The Democrats – they delay everything. They delay rocket launches. They are negative and hateful and have investigations and lawsuits against, you know, SpaceX and Tesla and even Neuralink.'

Why? She suspects darker forces at work. 'They put so many regulations in his way and they are very slow with approving anything. They just want to slow him down – and, you know, there are a lot of people behind that. They must be paying a lot to stop the development in America.'

Like his mother, Musk resents regulation – whether it's for rockets or for self-driving car software. 'I just think we've got far too many government agencies,' he says the day before the election. 'The federal bureaucracy has gotten out of hand, and we just need to pare it down to a sensible level.' His approach: slash first, fix later. 'If it turns out that, like, there's some regulation or agency that was doing something useful, we can put it right back. No problem. Like, it's like, "Oh, that regulation was important? No problem. We'll put it right back."'

It's a simple principle Musk applies to government: trial and error. But while this approach has its charms, there's also danger. Imagine the fallout if protections for drinking water, nuclear plants, or medical safety simply vanished. How many lives would be sacrificed to this 'let's just do it and see' logic?

Musk doesn't see it that way.

His compass is longtermism, the school of thought popularized by his favorite philosopher, William MacAskill. Present risk against future gain. What's the loss of millions today if billions can be saved later? For Musk, bad decisions are steps on the path to something greater. If no mistakes are made, not enough was tried. Tesla employees have told us how this mindset is drilled into them daily. Now Musk wants to scale it up – to a nation.

Even if states aren't companies, Musk wants to treat America's government like his factory floors: deliver or disappear. He wants to cut $2 trillion – 30 percent of the federal budget. It's a number that boggles the mind. How would that work? What would that look like?

'If Trump succeeds in forcing through mass deportations, combined with Elon hacking away at the government, firing people and reducing the deficit – there will be an initial severe overreaction in the economy,' one X user wrote just before the vote. 'Markets will tumble. But when the storm passes and everyone realizes we are on sounder footing, there will be a rapid recovery to a healthier, sustainable economy. History could be made in the coming two years.'

Musk's reply: 'Sounds about right.'

The billionaire is willing to accept sacrifice. More precisely: he's willing for Americans to sacrifice. The plan, he says at a rally, involves 'temporary hardship' for the average citizen. 'As a country, obviously, we need to live within our means.' He'll go through all government expenditures 'one item at a time, no exceptions, no special cases,' Musk says. 'There is so much government waste that it's kind of like being in a room full of targets, like you can't miss – you fire in any direction, you're going to hit a target.'

Many have promised to cut the national debt. Trump once pledged to erase it in eight years. Instead, it rose by nearly 8 trillion – a 39 percent jump – in half the time. Now both Trump and Musk are promising tax cuts – a key driver of the deficit. In October 2024, 23 Nobel Prize-winning economists warned that Trump's plans would mean 'higher prices, larger deficits, and greater inequality.'

And yet, does that matter? If Musk fails to deliver on his $2 trillion promise, his answer won't be a surprise to anyone who has followed his many promises in the past. Most likely, he will claim the figures were never meant to be taken literally. That they were just an exaggeration, an inflated fiction. Musk has tried this puffery defense successfully before, in court, with Tesla and Dogecoin. If damage is done, the blame lies with those who believed him.

When we began this book, we planned to end with a bleak outlook for Musk. To write about Tesla's sales problems, the job cuts, the courts calling Autopilot unfit for use. We wanted to show where Tesla was most vulnerable.

Then Musk made his move. Just when it looked like the rules might finally catch up with him, he set out to rewrite them.

Now he's arrived. A shadow president. Unelected. Unaccountable. But powerful beyond measure. He is re-shaping reality. He is inventing the future. He doesn't just want to define the next four years – but the next thousand. New technologies. A new humanity. No agency, no government, no investigator is meant to stand in his way. Musk is the one deciding what's right and what's wrong.

And so the richest man on earth has become what he never wanted to be: a politician.

The Reluctant Activist

Politics wasn't his thing, Musk said at a *Vanity Fair* event in 2015. He admitted that, because of his business interests, he couldn't entirely avoid it. But in principle, he said, he 'gets involved in politics as little as possible.'

For years, Musk stuck to that line – like many business leaders before him. He aimed for balance. In 2004, he donated $2,000 each to Republican president George W. Bush and his Democratic challenger John Kerry. Ahead of the 2008 election, he supported Barack Obama and Hillary

Clinton. In the 2016 race between Clinton and Donald Trump, Musk gave no money. 'I don't really have strong feelings, except that hopefully Trump doesn't get the nomination of the Republican party, because I think that's, yeah . . . that wouldn't be good,' Musk had said earlier. 'I think at most he would get the Republican nomination, but I think that would still be a bit embarrassing.'

When Trump became president, Musk's approach to political contributions shifted. He began supporting Republican candidates far more often than Democrats. He accepted Trump's offer to join his economic council, then resigned after the president pulled out of the Paris Agreement. 'Climate change is real,' Musk tweeted. 'Leaving Paris is not good for America or the world.'

When Trump lost the 2020 election, Musk said it would be better if the former president stayed out of politics. In summer 2022, he publicly backed Trump's Republican challenger Ron DeSantis and said Trump should go: 'I don't hate the man, but it's time for Trump to hang up his hat & sail into the sunset.'

Trump struck back. 'When Elon Musk came to the White House asking me for help on all of his many sub-sidized projects, whether it's electric cars that don't drive long enough, driverless cars that crash, or rocketships to nowhere, without which subsidies he'd be worthless and tell me how he was a big Trump fan and Republican, I could have said, "drop to your knees and beg," and he would have done it,' Trump wrote on his network Truth Social. Musk replied on Twitter: LMAO – short for Laughing my ass off.

At the time, Musk wasn't yet Twitter's owner. But he had already announced plans to buy the platform for $44 billion. His stated reason: concern for free speech. Musk accused Twitter of political bias in favor of the Democrats. In June 2022, he wrote: 'A platform cannot be considered inclusive or fair if it is biased against half the country.'

When Musk took over Twitter that October and fired nearly all employees responsible for moderating hate speech and disinformation, the platform, in the eyes of many, became a shithole. Major advertisers pulled out. Musk seemed to make a concession when he posted on October 27, 2022: 'Twitter obviously cannot become a free-for-all hell-scape, where anything can be said with no consequences!'

Three days later, Musk retweeted an article from the *Santa Monica Observer*, a website known for fake news. A conspiracy theorist had just broken into the home of Democratic politician Nancy Pelosi and attacked her husband with a hammer. The site falsely claimed Pelosi's husband had been drunk and fighting with a male prostitute. Musk shared the story with his more-than-100 million followers, writing: 'There is a tiny possibility there might be more to this story than meets the eye.'

In November 2022, Musk lifted Twitter's ban on Trump. The president had been removed after his supporters stormed the Capitol on January 6, 2021 – smashing windows, breaking down doors, and threatening to hang Vice President Mike Pence. Twitter's leadership feared Trump's posts might spark more violence. Musk called the ban a 'morally bad decision, to be clear, and foolish in the extreme.' Removing a sitting president, he said, had undermined public trust in Twitter for half the country.

According to Musk, X is meant to be a marketplace of ideas – a place where political discourse can unfold freely and without constraint. Studies suggest the opposite has happened. A report by the Center for Countering Digital Hate found that at least 87 of Musk's posts on the 2024 election contained false or misleading claims, reaching a total of 17 billion views. His posts alone reached twice as many people as all other political ads on X combined.

Since July 2024, engagement on Musk's posts has soared – views up 138 percent, retweets up 238 percent, likes up 186 percent. Many of the posts have supported Trump or cast Kamala Harris in a negative light. Interaction with Musk's content has far outpaced that of other political accounts, suggesting an algorithmic boost.

Musk treats transparency at X the same way he does at Tesla. Part of the algorithm remains secret, a black box – like the crash data in his electric cars. It allows Musk to go on praising his platform as the holy grail of free speech – without anyone able to verify it.

What's a President's Worth?

You don't need to be a programmer to detect Musk's partisanship. As reported by the *New York Times* and the *Wall Street Journal*, in April 2024 he meets with casino billionaire Steve Wynn, hedge fund manager Nelson Peltz, and other ultra-wealthy figures to discuss how best to help Republicans in the upcoming election. Shortly afterward, Musk throws himself into the campaign, launching a political action committee with a target budget of over $150 million, and sets out to recruit wealthy friends as donors and allies.

At Tesla's shareholder meeting, Musk says Trump sometimes calls him out of the blue – for no apparent reason. The two have spoken several times, he adds. When the former president survives an assassination attempt in July, Musk takes a public stance: 'I fully endorse President Trump and hope for his rapid recovery,' he writes on X to 200 million followers. 'Last time America had a candidate this tough was Theodore Roosevelt.'

Musk's role shifts quickly – from supporter to something better described as co-candidate. On August 12, ten weeks before the U.S. election, he hosts something on X that would have been unthinkable in 2020: a public conversation with Donald Trump. Musk has no operational role at X. He's not a manager, and no longer CEO. The idea that the owner of a media platform would personally appear alongside a presidential candidate feels surreal. No one at CNN, Fox News, or the *New York Times* would ever assign such a role to the owner rather than a reporter. In journalism, the wall between editorial and ownership is sacred. But here sits Musk – preparing to chat with Trump in person.

He insists it's nothing more than that. Not an interview – just a casual talk. 'I'm honored to have this conversation. I want to emphasize it's a conversation,' Musk says. 'And it's really intended to just get a feel for what Donald Trump is just like in a conversation.'

Maybe it's clever he puts it that way. Any journalist would be torn apart for admitting his only intent was to let a president speak unchecked. Musk preempts the criticism that he didn't challenge Trump hard enough. Then again, could there be a clearer admission? Musk says explicitly that he wants to offer Trump a platform. X becomes the mouthpiece of the former president. An echo chamber for the MAGA movement. Exactly the kind of campaigning Musk has long accused other media of providing.

After two hours, it's clear Musk has no journalistic ambition. He opens by asking Trump to describe his near-death experience at a rally in Pennsylvania a month earlier. 'I have to say that your actions at that assassination attempt were inspiring,' Musk says. The raised fist after the shots – exactly what he wants from a man who aims to lead the country. 'I think that is America. That is strength under fire. And so that's part of the reason why I was excited to endorse you as the President of the United States,' Musk tells Trump – and gives him room to ramble: 'But what was it like for you?'

It takes six minutes before Musk asks another question. 'Yeah, sure, absolutely,' he says as Trump speaks. The transcript shows Musk uttering 'yeah' 79 times. Whatever Trump claims, Musk agrees. None of Trump's many falsehoods are challenged. 'I have not been very political before,' Musk says. In the past, he mostly supported Democrats. Now he tells his audience: 'I think you should support Donald Trump for president. And I think it's actually a very important juncture in the road, and we're in deep trouble if it goes the other way.'

We're surprised that Musk's direct political intervention triggers no public outcry. Why is the reaction so different from just a few years ago, when other tech CEOs were under pressure for alleged political bias? In October 2020, Twitter CEO Jack Dorsey was summoned to the Senate – alongside the heads of Google and Facebook – to answer for the role of social media in spreading misinformation.

Back then, it was mostly Republicans who were outraged that Twitter dared to flag Trump's tweets when he spread falsehoods. 'Mr. Dorsey, who the hell elected you and put you in charge of what the media are allowed to report and what the American people are allowed to hear?' Republican senator Ted Cruz shouted. 'We're not doing that,' Dorsey replied. In March 2021, he and his colleagues were called back – this time mostly by Democrats, accusing the tech titans of letting dangerous misinformation about Covid-19 spread on their platforms.

The Point of No Return

Musk's open alignment with Donald Trump feels as though the debates over platform neutrality never happened. As though tech CEOs were never expected to remove false posts – or at least flag them. Again and again, the head of X spreads misinformation later debunked by media outlets or users – but by then, it's already been seen and shared millions of times. Musk seems to follow the old saying: 'A lie can travel halfway around the world while the truth is still putting on its pants.'

Musk throws himself into his new project with the same relentless drive that fueled PayPal, SpaceX, and Tesla: full commitment, no brakes, barely any sleep. Just as in the early days of his Twitter takeover, Musk surrounds himself with the people he trusts most. In the final stretch before election day, he enlists Steve Davis, CEO of the Boring Company, to assist Trump. He brings on Republican strategist Chris Young as political adviser. His PAC dispatches hundreds of field operatives to knock on doors and get out the vote. No one knows exactly how much Musk is spending. Official filings show close to $300 million. Trump, in private, puts the number at $500 million.

In their August conversation, Musk casually names the price of his support. 'I think it would be great to just have a government efficiency commission that ensures that the taxpayer money is spent in a good way,' he says to Trump. 'And I'd be happy to help out on such a commission.'

Days later, Reuters asks Trump if he's seriously considering Musk for a role. 'He's a very smart guy,' Trump replies. 'I certainly would, if he would do it; I certainly would. He's a brilliant guy.'

Later, Musk posts a photo: suit and tie, podium, American flag behind him. The lectern reads Department of Government Efficiency – DOGE. The same acronym as the joke cryptocurrency he once hyped, then mocked. His caption: 'I am willing to serve.'

What exactly does that mean?

Trump provides the answer on Fox News. When anchor Maria Bartiromo asks how he plans to cut the 'fat in government,' he points to his donor. 'I'm going to have Elon Musk,' Trump says. 'He is dying to do this [. . .] He's a great business guy actually, and he's a great cost-cutter.'

Trump describes the billionaire as a kind of miracle worker for the nation's finances. Musk, he says, told him he could cut costs 'without affecting anybody.' When Bartiromo asks if Musk would join his cabinet, Trump says: 'Not in the cabinet. He doesn't want to be in the cabinet. He wants to be in charge of cost-cutting, okay? We'll have a new position: Secretary of Cost-Cutting, okay? Elon wants to do that.'

Then he shifts tone – less policy, more persona. He marvels at the black cap Musk wore at one of his rallies, embroidered with the MAGA logo. 'He calls it Dark MAGA. How cool is that?' Trump says. 'He's actually campaigning. Because he says if we don't win, we're not going to have a country.'

In September, Musk warns that a Kamala Harris victory could mean 'the end of humanity.' SpaceX plans to send five unmanned ships to Mars over the next two years. If landings succeed, manned missions could follow within four more years. 'We want to enable anyone who wants to be a space traveler to go to Mars!' Musk posts on X. 'That means you or your family or friends – anyone who dreams of great adventure. Eventually, there will be thousands of Starships going to Mars and it will a glorious sight to see! Can you imagine? Wow.'

But there's a problem: the Democratic Party. Musk has long believed that humanity must become a 'multi-planetary species' in order to survive. Earth, in his view, is a ticking clock – one pandemic, one nuclear war, one catastrophe away from extinction. He sees himself as a pioneer of spacefaring civilization. And he feels unfairly held back.

'One of my biggest concerns right now is that the Starship programme is being smothered by a mountain of government bureaucracy that grows every year,' Musk writes. He says he has many concerns about Kamala Harris. But above all, he fears the red tape that would 'destroy the Mars program and doom humanity. It cannot happen.'

In early October, Musk speaks with Tucker Carlson. Once a star anchor at Fox News, Carlson had since become too toxic – even for the conservative network – after a string of extreme statements and conspiracy theories. The interview opens with a prediction: If Trump doesn't win, Musk is finished. 'If he loses, man, you're fucked,' Carlson says, laughing. Musk joins in: 'I'm fucked. If he loses, I'm fucked.'

They're in high spirits. Musk jokes about why no one's tried to kill the Democratic candidate. 'The Kamala puppet, I call her,' he says to Carlson. 'Nobody's even bothering to try to kill Kamala because it's pointless.' They laugh. 'What do you achieve? Nothing,' Musk adds. 'They'll just put in another puppet.'

'That's deep and true though,' Carlson says. Then Musk: 'Some people interpreted it as though I was calling for people to assassinate her, but I was like . . . Does it seem strange that no one's even bothered? Nobody tries to assassinate a puppet . . . She's safe.' They laugh again. 'It's not worth it,' Carlson says. 'There's an endless supply. It could be anybody.'

Both men are sure: if Trump loses, Musk is done. 'I can't deny it,' Musk says. 'I've been trashing Kamala non-stop.' Carlson replies, 'Oh, I know.' And they grin. 'How long do you think my prison sentence is going to be?' Musk asks. 'Will I see my children? I don't know.'

It sounds like a joke. But underneath the laughter, the message is clear. Musk is – the man who claimed he bought Twitter to protect public discourse – is all in for one candi-date, and one candidate only. And if you look closely, Musk is waging one of the dirtiest campaigns in modern history.

He posts a fake photo of the Democratic candidate in a hammer-and-sickle uniform – a symbol of communism. He shares deepfakes in which the multi-ethnic Harris calls herself incompetent and praises her candidacy as a win for diversity alone.

The Musk-funded PAC Future Coalition accuses Harris of favoring Jews or Muslims – depending on the audience. His money fuels a flood of fake ads on Facebook, falsely claiming that Trump's opponent wants to provide free healthcare for undocumented immigrants.

Days of Destiny

Shortly before the election, Musk sets up his headquarters in Pennsylvania – the state many experts believe will matter more than any other. On October 5, Trump appears in

Butler, the small town where he was shot at during a previous campaign rally. Then Trump introduces Musk. The billionaire leaps onstage, throws his arms in the air, and flashes his stomach.

'Donald Trump must win to preserve the Constitution,' Musk says. 'He must win to preserve democracy in America.'

Then Musk repeats what voters have heard so often from Trump: If Trump doesn't win, 'this will be the last election. That's my prediction. Nothing's more important,' Musk says. 'This is no ordinary election. The other side wants to take away your freedom of speech. They want to take away your right to bear arms. They want to take away your right to vote.'

In the final stretch of the campaign, Musk becomes a fixture in Pennsylvania. He joins Trump for appearance after appearance. He recycles long-debunked conspiracy theories – claims that mail-in ballots enabled fraud that was so sophisticated, it became untraceable.

He suggests the use of Dominion voting machines in Philadelphia is a bad omen. The company responds immediately: it isn't even active in Philadelphia. Multiple hand recounts in 2020 confirmed the accuracy of its machines. Back then, Dominion was a prime target for Trump and his supporters.

Musk shrugs this off. As election day nears, his methods grow increasingly bolder. 'If you're a registered Pennsylvania voter, you & whoever referred you will now get $100 for signing our petition in support of free speech & right to bear arms,' he posts on X on October 18. 'Earn money for supporting something you already believe in!'

In a livestream shortly after, Musk says: 'I'm here for a very important reason, which is – I can't emphasize this enough – Pennsylvania, I think, is the linchpin in this election. This election, I think, is going to decide the fate of America, and along with the fate of America, the fate of Western civilization.'

The next day, he raises the stakes – by a factor of 10,000. 'We are going to be awarding $1m randomly to people who have signed the petition,' Musk says at a town hall event in Pennsylvania. 'One of the challenges we're having is how do we get the public to know about this petition because the legacy media won't report on it.' Then he signs the first check and hands it to a lucky recipient.

Despite warnings from the Department of Justice that the offer could be illegal, Musk continues – and dominates headlines.

'Dramatically increasing my risk of being assassinated, and engaging in politics, are not what I want to do. I do not have a death wish,' he says during a campaign speech. Afterward, Musk posts the latest cover of Germany's primary news magazine, *Der Spiegel*. It shows a picture of Musk with the headline: 'Enemy Number Two'. Musk writes on X: 'With their relentless hit pieces, legacy mainstream media are actively encouraging the assassination of @realDonaldTrump and now me.'

It's a move straight from Trump's playbook: whatever others accuse him of, he throws back. Both men claim that hostile rhetoric from the other side incites violence. Both portray their opponents as a literal threat to life. According to Trump, Democrats are turning the U.S. into a 'garbage can.' According to Musk, a Harris victory wouldn't just end American democracy – it would end civilization.

On the eve of the election, Musk sits down with Joe Rogan. Rogan's podcast reaches 11 million listeners on average – more than any TV network. The conversation runs 2 hours and 38 minutes – a format unthinkable for traditional media. More than half the audience are men between 18 and 34 – a demographic most politicians struggle to reach.

Musk warns the country is about to end. 'If Trump doesn't win, this will be the last real election in America. If the big covenant Kamala puppet machine wins, they will legalize the illegals in the swing states. There'll be no swing states. Every election going forward will be a guaranteed Democrat win. And it'll actually be worse than California. The reason it'll be worse than California is because the one thing that keeps California from being super-crazy is that you can move outta California like you and I did. [. . .] But if, if Dems win this election, they will legalize enough illegals to turn the swing states and everywhere will be like California. There will be no escape.'

Rogan appears stunned, saying: 'That is so insane . . .'

Musk presses on: 'This is the final, this is it. This is the last chance.'

Rogan begins to ask: 'Has anyone tried to . . . ?' He doesn't finish. Musk turns to the camera, face serious: 'Go out and vote. Vote like your life depends on it. Vote like your future depends on it. Because it does. This is the last chance, man.'

On November 5, 2024, Americans cast their votes. More than 77 million choose Trump. For the first time in decades, a Republican not only wins more states – but also the popular vote. Fifty-six percent of young men cast their ballots for Trump.

Musk spends election night with Trump. Photos from the lavish party at Mar-a-Lago show the billionaire donor carrying his four-year-old son X on his shoulders through the crowd. Once Trump's victory is confirmed, his extended family gathers for a photo. Melania Trump is absent, but Musk poses with the Trump clan. The image is meme gold. Musk is dubbed the new First Lady and nick-named 'Elonia Trump.'

But the portrait of the new presidential family is more than just a viral joke. It's a proclamation. Trump is signaling how close Musk is to him – and how central the Tesla CEO is to his plans. And it's only the beginning.

The day after his historic victory, Trump receives congratulations from world leaders. One of them is Ukrainian President Volodymyr Zelensky, whose country has been locked in a brutal war since Russia's 2022 invasion. When Zelensky calls to congratulate him, Trump eventually passes the phone to his permanent guest: Elon Musk.

It's a chance to thank Musk for the Starlink satellite system. The moment likely leaves Zelensky deeply uneasy. According to the *Wall Street Journal*, Musk has been in 'regular contact' with Russian president Vladimir Putin since late 2022. They reportedly discuss 'geopolitics, business and personal matters.'

In the U.S., Musk is not just a supporter of the president – he is his partner.

In the three days following the election, Tesla's stock price climbs 28 percent. The company is now worth over $1 trillion. Musk becomes the first person with a net worth of $300 billion. A few weeks later, he crosses $400 billion.

He's looking at a future with unparalleled power. He commands a platform with more reach than any traditional media outlet in the world. With constant access to the White House, he can stymie the very regulators meant to oversee his companies. His satellite network can make armies see or be blind.

Whether Trump likes it or not, the most powerful man on the planet is Elon Musk.

In that light, 'The Tesla Files' may no longer be an issue for him. Yes, our investigation reveals how carelessly Tesla handles data from customers, employees, and partners. It shows how Musk reframes fatal Autopilot failures as signs of innovation. We've documented how he inflates Tesla's stock price with ever more fantastical promises that never materialize. Regulators, privacy watchdogs, and the SEC have all raised concerns about Musk's methods.

But what do concerns matter, if there's no one left to act on them?

Trump has already promised that one of his first acts in office will be to get rid of SEC chair Gary Gensler. He'll fire him 'on day one,' Trump says at a rally. On January 20, 2025, Gensler resigns. What successor would dare go after Trump's friend and partner Elon Musk – especially if Musk now holds the scissors to the agency's budget?

The same applies to the Federal Trade Commission (FTC), responsible for protecting consumer rights and ensuring fair competition. Musk has long been at odds with FTC chair Lina Khan. Her agency is investigating Tesla over misleading advertising of Autopilot. At X, the FTC is probing potential violations of data privacy law. And at SpaceX, it's reviewing reports of unfair labor practices – most notably the firing of critical employees.

Five days before the election, Musk posts: 'She will be fired soon.' In January 2025, Khan resigns.

The way disinformation and hate speech spread on Musk's platform X has raised concerns not only at the FTC, but also within the European Commission. Since 2022, EU law allows social networks to be sanctioned if they fail to protect users from illegal content and disinformation.

But the idea that Musk could actually face punishment leads Trump's running mate JD Vance to raise the stakes. In a September 2024 interview with YouTuber and ex-soldier Shawn Ryan, Vance says: 'If NATO wants us to continue supporting them, and NATO wants us to continue to be a good participant in this military alliance, why don't you respect American values and respect free speech?'

Freedom of speech is an American value. So is an independent judiciary, or it used to be. Like Trump, Musk supports investigations only when they don't target him or his allies.

Special counsel Jack Smith led two cases against Trump. In June 2023, he indicted him on 37 counts for mishandling classified documents, including 'willful retention of national defense information' and 'conspiracy to obstruct justice.' In August, Smith added four more counts, including 'conspiracy to defraud the United States' – for trying to overturn the 2020 election and Trump's role in the Capitol attack on January 6.

Trump has promised to fire Smith 'within two seconds' of taking office. Before the inauguration, Smith resigns.

But Musk isn't satisfied. On X, he posts: 'Jack Smith's abuse of the justice system cannot go unpunished.'

Outpaced by Reality

We're not Americans, but we're under no illusions either. Just before handing in the first manuscript for this book, we sit in the *Handelsblatt* cafeteria and ask ourselves: Does Musk's power have any limits?

We fear it doesn't. Even before his alliance with Trump, he had inserted himself into Canada's Covid protests and the Thai cave rescue – matters far beyond his domain. He supported Ukraine's defense against Russia with Starlink, but refused to support a Crimea offensive. On Taiwan, he speculated that it might become 'a special administrative zone' of China.

He made all these calls while insisting that politics didn't interest him. So what will he allow himself now, seated at the very center of power?

We think of Lord Acton – the British historian, politician, and moral philosopher. In 1887, he wrote: 'Power tends to corrupt, and absolute power corrupts absolutely. Great men are almost always bad men, even when they exercise influence and not authority.'

Musk already has the American president in his pocket.

What would stop him from doing to the world what he did to the U.S.?

A day later, we have our answer.

As Musk prepares for the next four years, Germany's coalition government collapses. The parties argue over when to hold new elections. The future is unclear. Musk, meanwhile, is certain.

On X, he posts: 'Olaf is a fool,' referring to German Chancellor Olaf Scholz. Then he calls Economy Minister Robert Habeck a fool too – for supporting tighter regulation of platforms like X.

German media respond with dry sarcasm. But it's not just a sideshow. It's a claim to authority. Musk speaks to 200 million people anytime he wants. On his platform, he defines what is true and what is false, who's a fool and who's fit to lead.

'Only the AfD can save Germany,' he writes on X in December 2024. When criticized, he writes an op-ed in the German newspaper *Welt am Sonntag*. 'To those who condemn the AfD as extremist, I say: don't be swayed by the label. Look at their policies, their economic plans, their commitment to cultural preservation. Germany needs a party unafraid to challenge the status quo.'

That the AfD opposed his Gigafactory in Grünheide doesn't seem to matter. On January 9, Musk hosts a live chat with AfD leader Alice Weidel – just as he did with

Trump. They bash the education system, curse red tape, and chat about whether Hitler was a leftist. The billionaire and the far-right leader hit it off.

On January 20, 2025, Musk speaks at an inauguration party in Washington, D.C., as Trump begins his second term. He places his hand on his heart. Then he lifts his right arm – smoothly, almost mechanically. The hand shoots up, fingers together, palm tilted downward. He bites his lip. His eyes sweep the crowd, as if to say: *Take that!*

In the days that follow, the world asks: Did Elon Musk just give the Nazi salute? He denies it. But when activists project the scene onto a wall of the Gigafactory in Grünheide, German authorities launch an investigation. The gesture itself – and its public display – is a criminal offense.

A few days later, Musk joins an AfD campaign event by video. He urges Germans to stop obsessing over guilt and start fighting for the future. The upcoming election, he says, is decisive – not just for Europe, but for the world, for civilization itself. He calls the current government totalitarian, claiming it suppresses free speech and targets dissent. That, he says, is dangerous. His wish: 'I hope Alice Weidel becomes chancellor.'

Musk also gets along great with Giorgia Meloni. Italy's right-wing prime minister calls Trump to congratulate him – then speaks with her 'friend Elon Musk' about 'a spirit of cooperation aimed at tackling future challenges,' as she posts on X. 'I'm convinced that his commitment and vision can be an important asset for both the United States and Italy.'

When a court blocks her party's plan to deport refugees to Egypt, Musk fumes. 'This is unacceptable,' he writes. 'Do the people of Italy live in a democracy or does an unelected autocracy make the decisions?' Days later, it's revealed that Meloni's government is negotiating a multi-billion-dollar deal with SpaceX for secure telecommunications.

Then things start to accelerate.

Turkish president Recep Tayyip Erdoğan calls Musk to propose a partnership. Argentina's Javier Milei, speaking at a Mar-a-Lago gala, offers to advise him on deregulation. Musk reportedly holds a secret meeting with Iran's UN ambassador to discuss easing tensions with the U.S.

He brands the German rescue NGO Sea-Watch a criminal organization. He declares that the Italian judges who blocked deportations 'must go.' He likens Britain's hate speech laws to Orwellian dystopias. He demands the release of Tommy Robinson – the UK's most notorious neo-Nazi – and calls for Prime Minister Keir Starmer to be jailed.

On January 6, 2025, he lets his followers vote on whether America should 'liberate the people of Britain from their tyrannical government.' Fifty-eight percent say yes.

At home, Musk's DOGE task force gains access to the Treasury Department's payment systems. A senior official who resisted resigns. Critics like Senator Ron Wyden warn of data abuse and politically motivated payment blocks.

These are the final days before our deadline, and we can barely keep up. Each new post from Musk makes us flinch. This is exactly the scenario we discussed just days ago in the *Handelsblatt* cafeteria. Reality hasn't caught up with us – it's raced ahead.

We're left asking: Where does this end?

Four years ago, tech CEOs were being yelled at in Senate hearings for allegedly influencing elections. Today, Musk shows how much the rules have changed. You no longer have to follow them. You don't even have to pretend. As in one of Musk's video games, the only goal is to win – by any means possible.

We sense a turning point. After the election, Trump's transition team proposes scrapping the reporting requirement for automated-driving accidents. Soon after his inauguration, as Musk flexes his new influence, operatives from his DOGE unit quietly sideline roughly 75 staff at the National Highway Traffic Safety Administration (NHTSA) across multiple divisions. The cuts include three of the seven lead vehicle-automation safety investigators tasked with probing Autopilot and full self-driving crashes. Their abrupt removal only comes to light via a Freedom of Information Act lawsuit by American Oversight, which reveals efforts to classify DOGE's records as 'presidential' and keep them under wraps for almost ten years.

In April, there's a breakthrough. The NHTSA rolls out a revised Standing General Order exempting Level 2 assisted-driving crashes that result only in tow-aways — cutting roughly 12 percent of Tesla's semi-autonomous reports since July 2021 and shielding low-severity Autopilot mishaps from scrutiny. Safety advocates warn this move hinders early detection of emerging system failures, leaving drivers and investigators 'in the dark,' while regulators defend it as a necessary trimming of 'regulatory burdens' to foster innovation. Tesla stands to avoid reporting hundreds of minor ADAS (Advanced Driver Assistance Systems) crashes each year — data that had previously fuelled calls for tighter oversight of its Autopilot and full self-driving software. Once again, Musk seems to get exactly what he paid for by funnelling hundreds of millions of dollars into Trump's campaign.

All this, the give-and-take between Musk and the new U.S. administration, plays out in the open. But as Musk wields his newfound power and his surrogates slash tens of thousands of government jobs, he faces a backlash. Tesla must announce global declines in sales.

The numbers are stark. In the first quarter of 2025, Tesla reports a net income of $409 million on revenues of $19.3 billion – its lowest profitability in years – trailing General Motors, Mercedes-Benz, Toyota, and nearly every other major automaker.

For the first time, it seems, the public is taking a truly hard look at Musk's company. Tesla's lineup is aging; the only updates are facelifts. The long-promised affordable model has vanished. The Cybertruck, once hailed as the greatest car ever built, falls far short of projections. And the Robotaxi – a revenue dream Musk has touted – already shows signs of the long development slog that has dogged so many of his ventures.

Dig deeper and the picture darkens. Tesla announces profitability is down 71 percent. That's bad, but it gets worse. The core automotive business isn't profitable at all.

An often-overlooked pillar has propped up Tesla's books: carbon-emission credits. Under global carbon-dioxide regulations, automakers that exceed fleet-wide targets earn credits; those that fall short must purchase them or pay fines. Tesla's all-electric fleet accumulates a glut of credits, which it then sells to legacy rivals.

That market has exploded – compliance-credit revenues soared from under $200 million in 2015 to over $3 billion in 2024 – making it one of Tesla's most lucrative revenue streams. Money flows straight from competitors' balance sheets into Tesla's, effectively subsidizing the core business and masking its losses.

It is an irony of history: the strictest regulators – California above all – have saved Musk from completely losing face. In the first quarter of 2025, Tesla booked $595 million from credit sales. Strip that out, and the $409 million profit turns into a substantial loss. Over 330,000 cars sold in the quarter – and not a cent of true automotive profit. Musk's relentless crusade against 'overbearing' regulations now rings hollow. After all his pioneering rhetoric, his most valuable company survives only on rules he claims to despise – and on the very regulators he vilifies.

None of this bodes well.

Tesla's core business is at the mercy of policy decisions – some in Musk's pocket, many out of reach in Brussels and Beijing. It is hardly a recipe for stability or sustainable growth. That same quarter, the stock price slides by more than 30 percent. With that, resentment grows.

'The company's reputation has just been destroyed by Elon Musk,' Ross Gerber tells *Sky News*. The wealth manager was an early investor in Tesla; now he's one of the first meaningful shareholders to call for Musk's resignation. 'It's time for somebody to run Tesla. The business has been neglected for too long. There are too many important things Tesla is doing, so either Elon should come back to Tesla and be the CEO of Tesla and give up his other jobs or he should focus on the government and keep doing what he is doing but find a suitable CEO of Tesla.'

How realistic is that scenario, with Musk still being the company's largest shareholder? Soon, media reports claim the board has begun searching for a successor. But, upon publication, the board denies it – and Musk vents, once again, about fake media. Then comes a major announcement: 'Starting next month, I'll be allocating far more of my time to Tesla,' Musk says on April 22. 'There are some challenges, and I expect that this year there'll probably be some unexpected bumps. But I remain extremely optimistic about the future of the company. The future of the company is fundamentally based on large-scale autonomous cars and vast numbers of autonomous humanoid robots.'

Musk goes on: 'I think Tesla will be the most valuable company in the world by far. It may be as valuable as the next five companies combined.' Again, he speaks of bumps in the road, but then rushes into his next prediction.

According to Musk, Tesla will 'be selling fully autonomous rides in June in Austin.' And that's just the start. By the second half of 2026, the CEO explains, Tesla's fully autonomous fleet will begin to have a significant impact on the financials of the company. 'It will really go exponential from there,' Musk says. People should look beyond the bumps, beyond the potholes of the road immediately ahead. And if they do, they will see a bright, shining light. 'I look forward to continuing to lead the team to great success in the future.'

The present is ugly: sagging sales, evaporating profits – and now outright violence. In Colorado, a woman hurls Molotov cocktails at Teslas and tags a dealership 'Nazi cars.' In South Carolina, police arrest a man who had set charging stations ablaze. In Oregon, a dealership's windows are smashed. In Seattle, four Cybertrucks go up in flames overnight. In Florida, a man defiles a Tesla pickup with feces.

The destruction spills over into Europe.

In Berlin, state security police launch an investigation after unknown perpetrators set fire to four Tesla vehicles. In the small northern town of Ottersberg, not far from Hamburg, seven cars are destroyed on the grounds of a dealership. In Toulouse, France, someone torches a dozen vehicles.

In the U.S., Musk once again sees himself as the victim of a political conspiracy. 'I thought the Democrats were supposed to be the party of empathy, the party of caring, and yet they're burning down cars and firing bullets into dealerships,' he says in an interview with Fox News. 'They basically want to kill me because I'm stopping their fraud. They want to hurt Tesla because we're stopping this terrible waste and corruption in the government and I guess they're bad people. Bad people will do bad things.'

Lucky for him, the U.S. president rushes to his aid. On March 10, 2025, Trump posts on his platform Truth Social: 'I'm going to buy a brand new Tesla tomorrow morning as a show of confidence and support for Elon Musk, a truly great American.'

The next day, Trump turns the South Lawn of the White House into a Tesla showroom.

Musk sends five vehicles, lining them up along the driveway for the president's personal inspection. 'It's very safe. It's very strong. Heavy. It's all steel, stainless steel,' Trump says of the Cybertruck, delivered with the batch. In the president's hands: a sheet of paper covered in hand-written notes – seemingly a sales pitch, complete with pricing details for various Tesla models.

White House officials livestream the event on Musk's social media platform X. It's a commercial straight from the Oval Office. 'They're harming a great American company,' Trump says, referring to the vandalism targeting Tesla. He calls Musk 'a patriot.' When asked by a reporter whether his event for Tesla will boost the company's sales and share price, he replies, 'I hope it does.'

But the president isn't content to leave things to hope – he wants to act. Asked whether protesters should be labelled 'domestic terrorists,' Trump responds: 'I'll do it.' These people, he says, were 'harming a great American company.' He then issued a warning: anyone using violence against Tesla would regret it. 'We're going to catch you and you're going to go through hell.'

At *Handelsblatt*, we watch in disbelief – unable to square this with anything we have seen before. However far we look back, we find no precedent for what's unfolding. Rockefeller, Ford, Morgan, Walton, the Koch brothers – all shaped politics in their time. But none of them ever installed a president. And then governed alongside him.

What now? What's to stop another tech titan from

backing a candidate, buying political power, and un-apologetically pushing through their own agenda? The early answers are already in: nothing.

In January, even before Trump and Musk officially take power, Facebook founder Mark Zuckerberg sends a video message to his 3.5 billion users worldwide:

'Hey everyone. I want to talk about something important today because it's time to get back to our roots,' Zuckerberg says, calling the U.S. elections 'a cultural tipping point towards, once again, prioritizing speech.' He continues: 'Here's what we're going to do. First, we're going to get rid of fact-checkers and replace them with community notes, similar to X.'

Soon after, Facebook's holding company Meta agrees to pay Trump $26 million in a legal dispute stemming from the aftermath of the 2020 elections. Meta banned Trump from its platforms, citing his election denial and role in the storming of the Capitol. According to Facebook, Trump 'actively fomented a violent insurrection designed to thwart the peaceful transition of power.'

Now, his accounts are reinstated – and Meta pays the incoming president a small fortune. Zuckerberg, just like Musk, was once appalled by Trump. Now, both are throwing money at him. Musk has normalized something that many might call corruption. And while some are stunned by Zuckerberg's reversal, he is not alone.

February 26, 2025. Jeff Bezos jumps on X: 'From now on, the Post will champion personal liberty and free markets. Opposing views can publish elsewhere.' The announcement comes two weeks after he spiked an endorsement of Kamala Harris – ending the *Post*'s 50-year tradition of backing presidential nominees. Opinion Editor David Shipley quits on cue. Bezos let him walk without hesitation. Then he says, a new editor will 'own this direction.'

The new direction, of course, is the way of Elon Musk. His fusion of corporate reach and partisan leverage is no longer an anomaly; it's fast becoming the industry standard. And Bezos, it seems, moves with the flow.

How times have changed. Bezos bought the *Washington Post* in 2013, a move hailed at the time as a potential lifeline for independent journalism. When Trump entered the White House in January 2017, Bezos responded with a flourish of defiance: the paper adopted a new slogan, 'Democracy Dies in Darkness.'

Then one of his journalists was killed.

On October 2, 2018, *Washington Post* columnist Jamal Khashoggi walked into the Saudi consulate in Istanbul and never came out. A 15-man hit squad was waiting for him. Inside, they suffocated him, dismembered his body with a bone saw, and made the remains disappear.

The CIA concluded – with 'high confidence' – that Saudi Crown Prince Mohammed bin Salman had personally ordered the assassination. The *Washington Post* responded with fury. Its editorial board ran blistering condemnations. Staffers wore his name on their badges. Bezos personally funded a private investigation into the leak of Khashoggi's texts.

Trump was less interested. 'Maybe he did and maybe he didn't!' the president said about the likelihood of the Saudi Crown Prince even knowing about the assassination before it happened. 'In any case, our relationship is with the Kingdom of Saudi Arabia.'

In 2020, Trump was voted out in an election he called rigged. When he returned as president in 2025, Bezos attended a high-end dinner at Mar-a-Lago and Amazon donated $1 million to the inauguration festivities. That was followed by an offer to pay $40 million for the rights to a documentary on Trump's wife, Melania.

Now, democracy was dying in bright daylight. In May 2025, Trump led a high-profile delegation of U.S. business leaders to Riyadh, the capital of Saudi Arabia. And at the U.S.–Saudi Investment Forum, Amazon announced a $5 billion partnership with Humain, Crown Prince Mohammed bin Salman's flagship AI initiative.

The murder of Jamal Khashoggi seemed but a distant memory.

Not to be outdone, Musk accompanied Trump on this swing through the Arab world in person, breezing through Riyadh like a one-man tech carnival. He clinched a Starlink license for aviation and maritime connectivity across the kingdom's skies and seas, cueing two Optimus robots to shimmy to 'YMCA' for Trump and the Crown Prince while aides inked the deal.

Musk strode through the petro-monarchies in a black T-shirt amid bespoke thobes, posing for pictures with the grin of a man who'd just bought the future. Again.

As Trump toured the Gulf, more than thirty U.S. business leaders gathered for a gilded lunch at the Saudi royal court – executives from Wall Street, Big Tech, defense contractors, and retail giants – all of them eager to secure their place in the new order.

And so, as Bezos and Zuckerberg and the others fall in line, the world is put on notice: The 2024 U.S. election may have been only the dress rehearsal – not the finale.

'If Trump loses, I'm fucked,' Musk admitted four weeks before the election, when he was in a bind: The polls showed Trump and Kamala Harris locked in a dead heat, uncertainty looming.

Musk knew one thing: He couldn't afford a Trump defeat. So he bought himself a Trump victory.

Now he gets to rule what he paid for. His credo: 'The world belongs to me; everyone else just lives in it.'

Our conclusion: We're all Musked.

Acknowledgements

This book would not have been possible without the colleagues who joined us in investigating the Tesla Files at *Handelsblatt*. Lars-Marten Nagel from our investigative team contributed his data expertise to help us make sense of the many gigabytes from Musk's empire. Co-head of investigations Martin Murphy was invaluable thanks to decades of industry contacts. Our teammates René Bender, Volker Votsmeier, and Vinzenz Neumaier were always ready to jump in when there was something to research.

Martin Kölling had a sixth sense when he unexpectedly received the first call from Krupski and passed it on to us. In New York, Felix Holtermann dove deep into Tesla's many U.S. legal battles. Astrid Dörner and Katharina Kort laid important groundwork in the U.S., while Stephan Scheuer sent in valuable insights from Tesla's home turf in Silicon Valley. Industry expert Thomas Jahn contributed his know-how and connections. Roman Tyborski coaxed childhood details out of Krupski's father in Polish. Trainee Martin Müller lent a hand wherever it was needed.

The *Tesla Files* would never have been published if we hadn't been able to rely on Peter Koppe. *Handelsblatt*'s chief legal counsel was an essential pillar during the groundwork for our reporting. It was also his idea to bring in Professor Roger Mann as an external legal advisor; his expertise and input were a huge asset. The same is true of the thorough analysis of the Tesla Files carried out by Martin Steinbach's team at the Fraunhofer Institute for Secure Information Technology, in the Media Security and IT Forensics division.

Behind every major story are names that don't appear in the byline. IT chief Yorn Zische helped create a secure environment for our research. With our colleagues Sven Prange and Christian Rickens, we had long discussions about structure and focus for our articles. Graphics chief Michel Becker brought his talent to the weekend editions we published in May and November 2023. None of this would have happened if our editor-in-chief, Sebastian Matthes, hadn't backed us from day one. It's true, he had his doubts at first. But without that healthy scepticism, critical journalism wouldn't be what it's supposed to be. A legal brawl with a litigious billionaire isn't something an editor-in-chief can take on lightly, let alone the potential pressure of agitated shareholders who might be looking for a scapegoat for billions in losses. Matthes had the backing of our managing director Andrea Wasmuth and our publisher Dieter von Holtzbrinck. We thank them as well.

First and foremost, though, we owe thanks to our agent, Thomas 'Tommy' Schmoll. He approached us in early November 2023 with the idea of a Tesla book. We politely declined – too much work, not enough time. But Tommy isn't someone you brush off easily. We lost count of how many times he half-urged, half-pleaded with us to take this project on. All we can say is: he won.

In the end, of course, it also took someone willing to publish our story in this form. Without C.H. Beck, you wouldn't be holding this book in your hands. Special thanks to Anke Hügler, Claire Zander, and Matthias Hansl.

We're especially grateful to Andrew Gordon and Dan Bunyard for believing that our story deserved to be read beyond Germany – and to Penguin Random House for having the courage to print it.

All the expertise and support would have meant nothing without the backing we had at home. We'd rather not count the hours we were physically with our families, but mentally still caught up in Elon Musk and Tesla. Sudden calls from sources on a Saturday at the soccer field, urgent messages on a Sunday night at dinner, half-nights spent on the laptop, sudden bursts of writing on holidays or en route to vacations – our loved ones had to put up with a lot. So, our deepest thanks go to our partners, Ayuk and Ronja, and to our children, Brandon, Dylan, Melinda, and Jelle. It can't have been easy, having to share attention with a global corporation. Then again, we think we heard a touch of pride in their voices when even the youngest said: 'Daddy's writing a book.'

S